Basics of Food Microbiology

Basics of Food Microbiology

Abhinit Kumar

RANDOM
RANDOM PUBLICATIONS
NEW DELHI (INDIA)

Contents

1

Food Microbiology: An Overview

Microbiology is the study of microscopic organisms, either unicellular (single cell), multicellular (cell colony), or acellular (lacking cells). Microbiology includes the disciplines virology, mycology, parasitology, bacteriology, and so on.

Eukaryotic microorganisms exhibit cell organelles and include fungi and protists, whereas prokaryotic organisms—which all are microorganisms—are conventionally classified as lacking organelles and include eubacteria and archaebacteria. Microbiologists traditionally relied on culture, staining, and microscopy. Apparently, however, only some 1% of the microorganisms present in some environments are culturable. Microbiologists often rely on extraction or detection of nucleic acid, either DNA or RNA sequences.

Viruses are not always classified as organisms, as they have been identified either as very simple microorganisms or very complex molecules. Prions, never considered microorganisms, have been investigated by virologists, however, as the clinical effects traced to them were originally presumed due to chronic viral infections, and virologists took search—discovering "infectious proteins".

As an application of microbiology, medical microbiology is often introduced with medical principles of immunology as microbiology and immunology. Otherwise, microbiology, virology, and immunology as basic sciences have greatly exceeded the medical variants, applied sciences.

History of Microbiology

The existence of microorganisms was hypothesized for many centuries before their actual discovery. The existence of unseen microbiological life was postulated by Jainism which is based on Mahavira's teachings as early as 6th century BCE. Paul Dundas notes that Mahavira asserted existence of unseen microbiological creatures living in earth, water, air and fire. Jain scriptures also describe nigodas which are sub-microscopic creatures living in large clusters and having a very short life and are said to pervade each and every part of the universe, even in tissues of plants and flesh of animals. The Roman Marcus Terentius Varro made references to microbes when he warned against locating a homestead in the vicinity of swamps "because there are bred certain minute creatures which cannot be seen by the eyes, which float in the air and enter the body through the mouth and nose and there by cause serious diseases."

In 1546 Girolamo Fracastoro proposed that epidemic diseases were caused by transferable seedlike entities that could transmit infection by direct or indirect contact, or vehicle transmission. However, early claims about the existence of microorganisms were speculative, and not based on microscopic observation. Actual observation and discovery of microbes had to await the invention of the microscope in the 17th century.

In 1676, Anton van Leeuwenhoek observed bacteria and other microorganisms, using a single-lens microscope of his own design. While Van Leeuwenhoek is often cited as the first to observe microbes, Robert Hooke made the first recorded microscopic observation, of the fruiting bodies of molds, in 1665. The first observation of microbes using a microscope is generally credited to the Dutch draper and haberdasher, Antonie van Leeuwenhoek, who lived for most of his life in Delft, Holland. It has, however, been suggested that a Jesuit priest called Athanasius Kircher was the first to observe micro-organisms. He was among the first to design magic lanterns for projection purposes, so he must have been well acquainted with the properties of lenses. One of his books contains a chapter in Latin, which reads in translation – 'Concerning the wonderful structure of things in nature, investigated by Microscope. Here, he wrote 'who would believe that vinegar and milk abound with an innumerable multitude of worms.' He also noted that putrid material is full of innumerable creeping animalcule. These observations antedate Robert Hooke's Micrographia by nearly 20 years and were published some 29 years before van Leeuwenhoek saw protozoa and 37 years before he described having seen bacteria.

The field of bacteriology (later a subdiscipline of microbiology) was founded in the 19th century by Ferdinand Cohn, a botanist whose studies on algae and photosynthetic bacteria led him to describe several bacteria including Bacillus and Beggiatoa. Cohn was also the first to formulate a scheme for the taxonomic classification of bacteria and discover spores. Louis Pasteur and Robert Koch were contemporaries of Cohn's and are often considered to be the father of microbiology and medical microbiology, respectively. Pasteur is most famous for his series of experiments designed to disprove the then widely held theory of spontaneous generation, thereby solidifying microbiology's identity as a biological science. Pasteur also designed methods for food preservation (pasteurization) and vaccines against several diseases such as anthrax, fowl cholera and rabies. Koch is best known for his contributions to the germ theory of disease, proving that specific diseases were caused by specific pathogenic micro-organisms. He developed a series of criteria that have become known as the Koch's postulates. Koch was one of the first scientists to focus on the isolation of bacteria in pure culture resulting in his description of several novel bacteria including Mycobacterium tuberculosis, the causative agent of tuberculosis.

While Pasteur and Koch are often considered the founders of microbiology, their work did not accurately reflect the true diversity of the microbial world because of their exclusive focus on micro-organisms having direct medical relevance. It was not until the late 19th century and the work of Martinus Beijerinck and Sergei Winogradsky, the founders of general microbiology (an older term encompassing aspects of microbial physiology, diversity and ecology), that the true breadth of microbiology was revealed. Beijerinck made two major contributions to microbiology: the discovery of viruses and the development of enrichment culture techniques. While his work on the Tobacco Mosaic Virus established the basic principles of virology, it was his development of enrichment culturing that had the most immediate impact on microbiology by allowing for the cultivation of a wide range of microbes with wildly different physiologies. Winogradsky was the first to develop the concept of chemolithotrophy and to thereby reveal the essential role played by micro-organisms in geochemical processes. He was responsible for the first isolation and description of both nitrifying and nitrogen-fixing bacteria. French-Canadian microbiologist Felix d'Herelle co-discovered bacteriophages and was one of the earliest applied microbiologists.

Characteristics of Microbes

A microorganism or microbe is a microscopic organism that comprises either a single cell, cell clusters, or multicellular relatively complex organisms. The study of microorganisms is called microbiology, a subject that began with Anton van Leeuwenhoek's discovery of microorganisms in 1675, using a microscope of his own design.

Microorganisms are very diverse; they include all of the prokaryotes, namely the bacteria and archaea; and various forms of eukaryote, comprising the protozoa, fungi, algae, microscopic plants (green algae), and animals such as rotifers and planarians. Some microbiologists also classify viruses as microorganisms, but others consider these as nonliving. Most microorganisms are unicellular (single-celled), but this is not universal, since some multicellular organisms are microscopic, while some unicellular protists and bacteria, like Thiomargarita namibiensis, are macroscopic and visible to the naked eye.

Microorganisms live in all parts of the biosphere where there is liquid water, including soil, hot springs, on the ocean floor, high in the atmosphere and deep inside rocks within the Earth's crust. Microorganisms are critical to nutrient recycling in ecosystems as they act as decomposers. As some microorganisms can fix nitrogen, they are a vital part of the nitrogen cycle, and recent studies indicate that airborne microbes may play a role in precipitation and weather.

Microbes are also exploited by people in biotechnology, both in traditional food and beverage preparation, and in modern technologies based on genetic engineering. However, pathogenic microbes are harmful, since they invade and grow within other organisms, causing diseases that kill humans, other animals and plants.

Evolution of Microbes

Single-celled microorganisms were the first forms of life to develop on Earth, approximately 3 billion–4 billion years ago. Further evolution was slow, and for about 3 billion years in the Precambrian eon, all organisms were microscopic. So, for most of the history of life on Earth the only forms of life were microorganisms. Bacteria, algae and fungi have been identified in amber that is 220 million years old, which shows that the morphology of microorganisms has changed little since the Triassic period.

Microorganisms tend to have a relatively rapid evolution. Most microorganisms can reproduce rapidly, and microbes such as bacteria can also freely exchange genes through conjugation, transformation and transduction, even between widely-divergent species. This horizontal gene transfer, coupled with a high mutation rate and many other means of genetic variation, allows microorganisms to swiftly evolve (via natural selection) to survive in new environments and respond to environmental stresses. This rapid evolution is important in medicine, as it has led to the recent development of 'super-bugs' — pathogenic bacteria that are resistant to modern antibiotics.

Classification and Structure

Microorganisms can be found almost anywhere in the taxonomic organization of life on the planet. Bacteria and archaea are almost always microscopic, while a number of eukaryotes are also microscopic, including most protists, some fungi, as well as some animals and plants. Viruses are generally regarded as not living and therefore are not microbes, although the field of microbiology also encompasses the study of viruses.

Prokaryotes

Prokaryotes are organisms that lack a cell nucleus and the other membrane bound organelles. They are almost always unicellular, although some species such as myxobacteria can aggregate into complex structures as part of their life cycle. Consisting of two domains, bacteria and archaea, the prokaryotes are the most diverse and abundant group of organisms on Earth and inhabit practically all environments where some liquid water is available and the temperature is below +140 °C. They are found in sea water, soil, air, animals' gastrointestinal tracts, hot springs and even deep beneath the Earth's crust in rocks. Practically all surfaces that have not been specially sterilized are covered by prokaryotes. The number of prokaryotes on Earth is estimated to be around five million trillion trillion, or 5 × 1030, accounting for at least half the biomass on Earth.

Bacteria

Almost all bacteria are invisible to the naked eye, with a few extremely rare exceptions, such as Thiomargarita namibiensis. They lack membrane-bound organelles, and can function and reproduce as individual cells, but often aggregate in multicellular colonies. Their genome is usually a single loop of DNA, although they can also harbor small pieces of DNA called plasmids. These plasmids can be transferred between cells through bacterial conjugation. Bacteria are surrounded by a cell wall, which provides strength and rigidity to their cells. They reproduce by binary fission or sometimes by budding, but do not undergo sexual reproduction. Some species form extraordinarily resilient spores, but for bacteria this is a mechanism for survival, not reproduction. Under optimal conditions bacteria can grow extremely rapidly and can double as quickly as every 10 minutes.

Archaea

Archaea are also single-celled organisms that lack nuclei. In the past, the differences between bacteria and archaea were not recognised and archaea were classified with bacteria as part of the kingdom Monera. However, in 1990 the microbiologist Carl Woese proposed the three-domain system that divided living things into bacteria, archaea and eukaryotes. Archaea differ from bacteria in both their genetics and biochemistry. For example, while bacterial cell membranes are made from phosphoglycerides with ester bonds, archaean membranes are made of ether lipids.

Archaea were originally described in extreme environments, such as hot springs, but have since been found in all types of habitats. Only now are scientists beginning to realize how common archaea are in the environment, with crenarchaeota being the most common form of life in the ocean, dominating ecosystems below 150 m in depth. These organisms are also common in soil and play a vital role in ammonia oxidation.

Eukaryotes

Most living things that are visible to the naked eye in their adult form are eukaryotes, including humans. However, a large number of eukaryotes are also microorganisms. Unlike bacteria and archaea, eukaryotes contain organelles such as the cell nucleus, the Golgi apparatus and mitochondria in their cells. The nucleus is an organelle that houses the DNA that makes up a cell's genome. DNA itself is arranged in complex chromosomes. Mitochondria are organelles vital in metabolism as they are the site of the citric acid cycle and oxidative phosphorylation. They evolved from symbiotic bacteria and retain a remnant genome. Like bacteria, plant cells have cell walls, and contain organelles such as chloroplasts in addition to the organelles in other eukaryotes. Chloroplasts produce energy from light by photosynthesis, and were also originally symbiotic bacteria.

Unicellular eukaryotes are those eukaryotic organisms that consist of a single cell throughout their life cycle. This qualification is significant since most multicellular eukaryotes consist of a single cell called a zygote at the beginning of their life cycles. Microbial eukaryotes can be either haploid or diploid, and some organisms have multiple cell nuclei. However, not all microorganisms are unicellular as some microscopic eukaryotes are made from multiple cells.

Protists

Of eukaryotic groups, the protists are most commonly unicellular and microscopic. This is a highly diverse group of organisms that are not easy to classify. Several algae species are multicellular protists, and slime molds have unique life cycles that involve switching between unicellular, colonial, and multicellular forms. The number of species of protozoa is uncertain, since we may have identified only a small proportion of the diversity in this group of organisms.

Animals

Most animals are multicellular, but some are too small to be seen by the naked eye. Microscopic arthropods include dust mites and spider mites. Microscopic crustaceans include copepods and the cladocera, while many nematodes are too small to be seen with the naked eye. Another particularly common group of microscopic animals are the rotifers, which are filter feeders that are usually found in fresh water. Micro-animals reproduce both sexually and asexually and may reach new habitats as eggs that survive harsh environments that would kill the adult animal. However, some simple animals, such as rotifers and nematodes, can dry out completely and remain dormant for long periods of time.

Fungi

The fungi have several unicellular species, such as baker's yeast (Saccharomyces cerevisiae) and fission yeast (Schizosaccharomyces pombe). Some fungi, such as the pathogenic yeast Candida albicans, can undergo phenotypic switching and grow as single cells in some environments, and filamentous hyphae in others. Fungi reproduce both asexually, by budding or binary fission, as well by producing spores, which are called conidia when produced asexually, or basidiospores when produced sexually.

Plants

The green algae are a large group of photosynthetic eukaryotes that include many microscopic organisms. Although some green algae are classified as protists, others such as charophyta are classified with embryophyte plants, which are the most familiar group of land plants. Algae can grow as single cells, or in long chains of cells. The green algae include unicellular and colonial flagellates, usually but not always with two flagella per cell, as well as various colonial, coccoid, and filamentous forms. In the Charales, which are the algae most closely related to higher plants, cells differentiate into several distinct tissues within the organism. There are about 6000 species of green algae.

Habitats and Ecology

Microorganisms are found in almost every habitat present in nature. Even in hostile environments such as the poles, deserts, geysers, rocks, and the deep sea. Some types of microorganisms have adapted to the extreme conditions and sustained colonies; these organisms are known as extremophiles. Extremophiles have been isolated from rocks as much as 7 kilometres below the Earth's surface, and it has been suggested that the amount of living organisms below the Earth's surface may be comparable with the amount of life on or above the surface. Extremophiles have been known to survive for a prolonged time in a vacuum, and can be highly resistant to radiation, which may even allow them to survive in space. Many types of microorganisms have intimate symbiotic relationships with other larger organisms; some of which are mutually beneficial (mutualism), while others can be damaging to the host organism (parasitism). If microorganisms can cause disease in a host they are known as pathogens.

Extremophiles

Extremophiles are microorganisms that have adapted so that they can survive and even thrive in conditions that are normally fatal to most life-forms. For example, some species have been found in the following extreme environments:

— Temperature: as high as 130 °C (266 °F), as low as -17 °C (1 °F)
— Acidity/alkalinity: less than pH 0, up to pH 11.5
— Salinity: up to saturation
— Pressure: up to 1,000-2,000 atm, down to 0 atm (e.g. vacuum of space)
— Radiation: up to 5kGy

Extremophiles are significant in different ways. They extend terrestrial life into much of the Earth's hydrosphere, crust and atmosphere, their specific evolutionary adaptation mechanisms to their extreme environment can be exploited in bio-technology, and their very existence under such extreme conditions increases the potential for extraterrestrial life.

Soil Microbes

The nitrogen cycle in soils depends on the fixation of atmospheric nitrogen. One way this can occur is in the nodules in the roots of legumes that contain symbiotic bacteria of the genera Rhizobium, Mesorhizobium, Sinorhizobium, Bradyrhizobium, and Azorhizobium.

Symbiotic Microbes

Symbiotic microbes such as fungi and algac form an association in lichen. Certain fungi form mycorrhizal symbioses with trees that increase the supply of nutrients to the tree.

Benefits Microbes

While some fear microbes due to the association of some microbes with various human illnesses, many microbes are also responsible for numerous beneficial processes such as industrial fermentation (e.g. the production of alcohol, vinegar and dairy products), antibiotic production and as vehicles for cloning in more complex organisms such as plants. Scientists have also exploited their knowledge of microbes to produce biotechnologically important enzymes such as Taq polymerase, reporter genes for use in other genetic systems and novel molecular biology techniques such as the yeast two-hybrid system.

Bacteria can be used for the industrial production of amino acids. Corynebacterium glutamicum is one of the most important bacterial species with an annual production of more than two million tons of amino acids, mainly L-glutamate and L-lysine.

A variety of biopolymers, such as polysaccharides, polyesters, and polyamides, are produced by microorganisms. Microorganisms are used for the biotechnological production of biopolymers with tailored properties suitable for high-value medical application such as tissue engineering and drug delivery. Microorganisms are used for the biosynthesis of xanthan, alginate, cellulose, cyanophycin, poly(gamma-glutamic acid), levan, hyaluronic acid, organic acids, oligosaccharides and polysaccharide, and polyhydroxyalkanoates.

Microorganisms are beneficial for microbial biodegradation or bioremediation of domestic, agricultural and industrial wastes and subsurface pollution in soils, sediments and marine environments. The ability of each microorganism to degrade toxic waste depends on the nature of each contaminant. Since sites typically have multiple pollutant types, the most effective approach to microbial biodegradation is to use a mixture of bacterial species and strains, each specific to the biodegradation of one or more types of contaminants.

There is considerable evidence supporting microbial contributions to human and animal health. Some benefit may be be conferred by consuming fermented foods, probiotics (bacteria potentially beneficial to the digestive system) and/or prebiotics (substances consumed to promote the growth of probiotic microorganisms). The importance of bacteria to human health is only beginning to be investigated, a field that is certain to expand in light of recent discoveries that, in humans, bacterial cells outnumber somatic cells by a factor of 10 to one.

Recent research has suggested that microorganisms could be useful in the treatment of cancer. Various strains of non-pathogenic clostridia can infiltrate and replicate within solid tumors. Clostridial vectors can be safely administered and their potential to deliver therapeutic proteins has been demonstrated in a variety of preclinical models.

Use of Microbes in Food Production

Food microbiology is a subdiscipline in the field of microbiology concerned with the study of bacteria, fungi, and viruses that grow in or are transmitted by foods. While bacteria are frequently associated with food spoilage and food poisoning, some species preserve foods through fermentation or produce food ingredients. Food microbiology is a broad field that can include not only microbiology but also sanitation, epidemiology, biochemistry, engineering, statistics, and mathematical modelling.

When certain bacteria grow in foods, they produce desirable flavours and textures, and may also inhibit pathogenic organisms. Most of these bacteria belong to the genera Streptococcus, Lactobacillus, Leuconostoc, Pediococcus, or Micrococcus. They are used to make fermented dairy products, meats, and vegetables, and to preserve food by converting the sugars needed by competing microbes to lactic acid, which inhibits their growth. Acetobacter and Gluconobacter are used in the production of vinegar. Yeasts, usually Saccharomyces, which produce ethanol and carbon dioxide, are used in the processes of brewing and baking.

They also help to control population levels of higher animals and plants by parasitism and pathogenicity. When plants and animals die, their protective antimicrobial systems cease to function so that, sooner or later, decay begins liberating the smaller molecules for reuse by plants. Without human intervention, growth, death, decay and regrowth would form an intricate web of plants, animals and microorganisms, varying with changes in climate and often showing apparently chaotic fluctuations in population of individual species, but inherently balanced in numbers between producing, consuming and recycling groups.

In the distant past, these cycles of growth and decay would have been little influenced by the small human population that could be supported by the hunting and gathering of food. From around 10000 BC however, the deliberate cultivation of plants and herding of animals started in some areas of the world. The increased productivity of the land and the improved nutrition that resulted led to population growth and a probable increase in the average life span. The availability of food surpluses also liberated some from daily toil in the fields and stimulated the development of specialised crafts, urban centres, and trade—in short, civilisation.

In the dairy industry, many products result from fermentation by microorganisms in milk and the products of milk. For example, buttermilk results from the souring of low-fat milk by lactic acid. The flavor is due to substances such as diacetyl and acetaldehyde, which are produced by species of Streptococcus, Leuconostoc, and Lactobacillus as they grow.

A fermented milk product with a puddinglike consistency is yogurt. Two bacteria, Streptococcus thermophilus and Lactobacillus bulgaricus, are essential to its production. After the milk has been heated to achieve evaporation, the bacteria are added, and the condensed milk is set aside at a warm temperature to produce the yogurt. Sour cream is produced in a similar way, using cream as a starter material.

The protein portion of the milk, the casein, is used to produce cheese and cheese products. Precipitated from the milk, the protein curd is an unripened cheese such as cottage cheese. The leftover liquid, the whey, can be used to make cheese foods.

When the cheese is allowed to ripen through the activity of various microorganisms, various cheeses are produced. Soft cheeses, such as Camembert, do not spoil rapidly. Camembert cheese is a product of the growth of the fungus Penicillium camemberti. Hard cheeses have less water and are ripened with bacteria or fungi. Swiss cheese is ripened by various bacteria, including species of Propionibacterium, which produces gas holes in the cheese. Bleu cheese is produced by Penicillium roqueforti, which produces veins within the cheese as it grows.

Other fermented foods are also the product of microbial action. Sauerkraut, for example is produced by Leuconostoc and Lactobacillus species growing within shredded cabbage. Cucumbers are fermented by these same microorganisms to produce pickles.

Bread is still another product of microbial action. Flour, water, salt, and yeast are used to make the dough. The yeast most often used is Saccharomyces cerevisiae. This organism ferments the carbohydrates in the dough and produces carbon dioxide, which causes the dough to rise and creates the soft texture of bread. Unleavened bread is bread that contains no yeast. Sourdough bread can be made by using lactic acid bacteria to contribute a sour flavor to the doug

Microorganisms and Food Spoilage

Food is considered contaminated when unwanted microorganisms are present. Most of the time the contamination is natural, but sometimes it is artificial. Natural contamination occurs when microorganisms attach themselves to foods while the foods are in their growing stages. For instance, fruits are often contaminated with yeasts because yeasts ferment the carbohydrates in fruits. Artificial contamination occurs when food is handled or processed, such as when fecal bacteria enter food through improper handling procedures.

Food spoilage is a disagreeable change or departure from the food's normal state. Such a change can be detected with the senses of smell, taste, touch, or vision. Changes occurring in food depend upon the composition of food and the microorganisms present in it and result from chemical reactions relating to the metabolic activities of microorganisms as they grow in the food.

Various physical, chemical, and biological factors play contributing roles in spoilage. For instance, microorganisms that break down fats grow in sweet butter (unsalted butter) and cause a type of spoilage called rancidity. Certain types of fungi and bacteria fall into this category. Species of the Gram-negative bacterial rod Pseudomonas are major causes of rancidity. The microorganisms break down the fats in butter to produce glycerol and acids, both of which are responsible for the smell and taste of rancid butter.

Another example occurs in meat, which is primarily protein. Bacteria able to digest protein (proteolytic bacteria) break down the protein in meat and release odoriferous products such as putrescine and cadaverine. Chemical products such as these result from the incomplete utilization of the amino acids in the protein.

Food spoilage can also result in a sour taste. If milk is kept too long, for example, it will sour. In this case, bacteria that have survived pasteurization grow in the milk and produce acid from the carbohydrate lactose in it. The spoilage will occur more rapidly if the milk is held at room temperature than if refrigerated. The sour taste is due to the presence of lactic acid, acetic acid, butyric acid, and other food acids.

The general sources of food spoilage microorganisms are the air, soil, sewage, and animal wastes. Microorganisms clinging to foods grown in the ground are potential spoilers of the food. Meats and fish products are contaminated by bacteria from the animal's internal organs, skin, and feet. Meat is rapidly contaminated when it is ground for hamburger or sausage because the bacteria normally present on the outside of the meat move into the chopped meat where there are many air pockets and a rich supply of moisture. Fish tissues are contaminated more readily than meat because they are of a looser consistency and are easily penetrated.

Canned foods are sterilized before being placed on the grocery shelf, but if the sterilization has been unsuccessful, contamination or food spoilage may occur. Swollen cans usually contain gas produced by members of the genus Clostridium. Sour spoilage without gas is commonly due to members of the genus Bacillus. This type of spoilage is called flat-sour spoilage. Lactobacilli are responsible for acid spoilage when they break down the carbohydrates in foods and produce detectable amounts of acid.

Among the important criteria determining the type of spoilage are the nature of the food preserved, the length of time before it is consumed, and the handling methods needed to process the foods. Various criteria determine which preservation methods are used.

References

Diaz. E., 2008. *Microbial Biodegradation: Genomics and Molecular Biology* (1st ed.). Caister Academic Press.

Keen E.C., 2012. "Felix d'Herelle and Our Microbial Future".*Future Microbiology* 7 (12): 1337–1339.

Madigan M., Martinko J. 2006. *Brock Biology of Microorganisms* (13th ed.). Pearson Education.

Ryan KJ, Ray C.G., 2004. *Sherris Medical Microbiology* (4th ed.). McGraw Hill.

2

Use of Microbes in Food Production

Microbes are found throughout the world, in soil, on water, plants, animals, rocks, and people. After death, all living organisms decompose to their base elements of water, carbon, nitrogen, phosphate, and trace elements. This process is called bioremediation or mineralization. It takes approximately one million bacteria to recycle the 6 million organic molecules that make up life.The average human being has 3 lbs. of microbes on and in their bodies. It is difficult to describe a simple yet complex form of life like the microbes. They're so small that a powerful microscope is required to see them. It took the development of the microscope before man learned that microbes were the chemical agents of decay, fermentation, and disease. Their size makes the microbe the smallest living unit that can contain the necessary complex chemicals for life processes and the necessary enzymes for recycling complex organic matter. These small cells also have a large surface to volume ratio allowing for a maximum cell wall activity and interchange of materials in and out of the cells. In this way, nature fabricated the total microbe population for maximum activity in recycling of all natural organic matter.

Most microorganisms are too small to be seen, their importance cannot be ignored. Microorganisms are the foundation of the biosphere-both from an evolutionary and an ecological perspective. Microorganisms were the first organisms on Earth; they have lived on this planet for a period of at least 3.7 billion years of the 4.6 billion-year existence of the Earth. Microorganisms were living inhabitants for more than 3.0 billion years before the appearance of plants and animals. Not only did plants and animals evolve rather recently in Earth's history, but they evolved from microbial ancestors. A recent report of evidence for microbial life on Mars also is consistent with the concept that microorganisms preceded plants and animals on Earth.

The Earth's biosphere is largely shaped by geochemical activities of microorganisms that have provided conditions both for the evolution of plants and animals and for the continuation of all life on Earth. Many microorganisms carry out unique geochemical processes critical to the operation of the biosphere. Therefore, it is not surprising that the diversity of microorganisms-from genetic, metabolic, and physiological aspects-is far greater than that found in plants and animals. In contrast to plants and animals, the diversity of the microbial world is largely unknown, and, of that which is known, the diversity is spectacular.

Some microorganisms live at boiling temperatures, or higher, in hot springs and deep sea thermal vents; others live at temperatures below freezing in sea ice. Some produce sulfuric and nitric acids. Many grow without oxygen; the anaerobic activities of these microorganisms are necessary for carrying out the many essential processes in the environment that cannot be accomplished by plants and animals, including methane production and nitrogen fixation. Such familiar activities as leavening bread and production of yogurt, pickles, wine, beer, and cheeses rely on microorganisms carrying out the key processes.

Microorganisms are the foundation of the biosphere. Without them, other life forms would not have evolved and could not exist. Microorganisms established the geochemical conditions on Earth that enabled evolution of plants and animals. Plants and animals are descended from microorganisms, and their cells are now known to be composites of microorganisms. For example, the mitochondria of all plants and animals are derived from bacteria. Similarly, the photosynthetic organelle, the chloroplast, found in all plants and algae is descended from a group of photosynthetic bacteria, the cyanobacteria.

Cyanobacteria are believed to be the first organisms on Earth to produce free oxygen gas and, concomitantly, the protective ozone layer around Earth, thereby providing conditions for evolution of land plants and animals. Humans and other animals, as well as plants, are completely dependent on microorganisms for life. Like all animals, humans harbor billions of microorganisms in their digestive tracts, microorganisms necessary to digest food and provide nutrients, such as vitamins and amino acids, for growth and a source of energy. Plants also require microorganisms to provide nutrients for growth, an activity that takes place largely in root systems. There the organic materials in soil are broken down by bacteria and fungi to provide inorganic materials, such as nitrogen and phosphorus, the natural fertilizers made available by microorganisms and required by plants for growth and development.

Microorganisms exist everywhere physical conditions permit. Although lake water may appear transparent to the eye, a liter of the water can harbor a billion bacteria. A gram of soil can also contain over a billion bacteria. Many microorganisms have special dispersal cells that can be carried by winds across and between continents. In addition, birds and insects transport microorganisms as they fly. Thus, we live in a world teeming with microbial life that carries out a myriad of activities essential for sustaining the biosphere of Earth.

Microorganisms are highly diverse genetically and metabolically, far more so than plants and animals. This should not seem surprising because microorganisms have existed on Earth for over 3.5 billion years, whereas multicellular plants and animals have existed only 600 million years. From analyses of molecular sequences of genes, such as 16S and 18S ribosomal RNA, approximately 20 separate, main phylogenetic groups of microbial life have been identified, comparable in depth and breadth to the animal and plant kingdoms. Furthermore, microbiologists have discovered groups that represent new phyla, such as the Korarchaeota, not yet studied in pure culture.

One of the most surprising characteristics of microorganisms is the range of physiological conditions under which they flourish. They grow across broad ranges of temperature, pH, salt concentration, and oxygen concentration. Some thrive at boiling temperatures in hot springs and at temperatures higher than 100°C in submarine vents. Others are found in sea ice off Antarctica and at the North Pole. Some produce sulfuric and nitric acids, and many microbial species live without oxygen. Others live in saturated salt brines, and some are resistant to high levels of radioactivity.

The variety of metabolic types of microorganisms is enormous. Some are photosynthetic and, like plants, produce oxygen in this process. In fact, this "biotechnology" first occurred in the cyanobacteria, which subsequently evolved endosymbiotically to form chloroplasts that enable algae and plants to conduct photosynthesis. Other bacterial groups carry out photosynthesis by different pathways and produce products such as sulfur. Microorganisms are the primary, if not sole, agents responsible for degradation of a great variety of organic compounds, including cellulose, hemi-cellulose, lignin, and chitin (the most abundant organic matter on Earth). If it were not for microbial activities involved in natural decay, excessive amounts of organic matter would accumulate in forests and aquatic sediments.

In addition, microorganisms are responsible for degradation of toxic chemicals derived from anthropogenic sources, such as PCBs (polychlorinated biphenyls), dioxins and other pesticides. Because microorganisms are so versatile, they are relied upon to digest wastes in sewage treatment plants, landfills, and toxic waste sites. It is in this regard that the field of bioremediation, encompassing all of these processes, is still in its infancy. Much needs to be learned before microbial breakdown processes can be controlled and enhanced in situ. Microorganisms play important roles in geochemical processes. For example, the global nitrogen cycle in nature is dependent on microorganisms.

Unique processes carried out by microorganisms include nitrogen fixation (the natural conversion of atmospheric dinitrogen gas to utilizable organic cell nitrogen), oxidation of ammonia and nitrite to nitrate, and nitrate reduction with formation of dinitrogen and nitrous oxide gases. Similar important and unique roles are played in other cycles, such as the sulfur and carbon cycles, as well as in the oxidation and reduction of metals. If it were not for microorganisms, substances such as cellulose and lignin would not be recycled; they would accumulate in the environment. Indeed, almost all organic substances are recycled via activities of bacteria, fungi, and protozoa.

The importance of microorganisms in agriculture is enormous and extends beyond geochemical cycles. Indeed, most of the fertility of soil is derived from microbial mineralization and in production of nitrogen for plant growth. These processes extend to lichen- and cyanobacterial-dominated soils which occupy a larger surface area on Earth than in tropical rain forests. Mycorrhizal fungi form important rhizosphere associations with almost all plants. Such associations are essential for optimum growth and, in fact, permit some plants to grow in areas they could not otherwise colonize.

Recent advances in agriculture stem from breakthroughs in the genetic engineering of plants; one of the most dramatic examples is that of the bacterium Agrobacterium tumefaciens. Normally the causative of crown gall disease in plants, this bacterium has been used to transfer favorable properties into an agriculturally important plant species, thereby providing a mechanism for introducing genes that provide resistance to plant diseases, insects or pesticides into plants. Microorganisms are important in recycling waste materials. Sewage (wastewater) treatment and the breakdown of garbage in landfills occur because of microorganisms. These microorganisms do this "for free" because, in most cases, they derive energy from the process.

A recent discovery indicates that microorganisms may influence weather. Some marine algae produce dimethyl sulfide (DMS). This compound is volatile and escapes into the atmosphere where it is photo-oxidized to form sulfate. The sulfate acts as a water nucleating agent and when enough sulfate is formed, clouds are produced; these clouds have three major impacts. First, they shade the ocean and, thereby, slow further algal growth and DMS production, eventually decreasing cloud formation. Second, the clouds lead to increased rainfall. And third, because clouds are reflective of incoming sunlight, the clouds reduce the amount of heat that reaches Earth, moderating global warming.

Microorganisms are at the core of biotechnology. Many antibiotics and anti-tumor agents are derived from microorganisms, including penicillin, streptomycin, and chloramphenicol. The emergence of multiple antibiotic-resistant pathogenic bacteria has necessitated the search for new antibiotics. Because there are so many types of microorganisms, they produce many unique products currently useful in biotechnology and offer great promise for exploitation in the future.

Microbes in Industrial Food Production

Microbes and food have always had a love-hate relationship. Food producers have learned to harness their power — and sometimes for good use. There is no doubt that organically grown fruits and vegetables are superior in flavor. Now, despite arguments to the contrary made by food manufacturers, they may well be more nutritious for a variety of reasons.

There are always studies available to prove any point, especially one where chemically doused foods are more nutritious. But then, like bacteria in the throes of death, food manufacturers mount a valiant effort in trying to convince those near-at-hand that what they say is valid. But, be that as it may, the choice is ultimately made individually.

Bacteria are, to a great extent, resistant to the effects of irradiation because of their ability to repair their damaged DNA. Sometimes, however, their repair is faulty, and the new DNA becomes abnormal, causing a new strain. *Salmonella* and *Campylobacter* are easily removed by irridation, but other bacteria are more stubborn. Some, like *Clostridium botulinum* and *L. monocytogenes*, have the ability to survive the drying techniques of irridation by converting themselves into tough, little spheres.

Food irriadiation, however, exposes an edible product to the equivalent of two and a half million chest X-rays. How many nutrients are left after this bombardment is anyone's guess, but, at least, most of the germs are gone. Most, not all. Short term studies have shown the destruction of numerous vitamins through this process as well as the formation of new carcinogens. There is also the possible formation of mutant bacteria and viruses. Since the USDA approved such a process in February of 2000, fast food chains, restaurants, school lunch programs, and cafeterias are now using irradiated meat products without the public knowledge.

The most common usage for bacteria in food preparation is with dairy fermentations. Yogurt and cheeses have been made for centuries using bacteria. The ancients may not have known exactly what kind of bacteria that was needed or if what was needed was, indeed, bacteria. All they knew was that the previous batch was required to make a new one.

Many people lack the ability to break down and absorb lactose, the sugar molecule in milk. As a result, it enters the gut, producing acid and gas, causing pain and diarrhea. Fermented milk products metabolize lactose into lactic acid, which is more tolerable for many people. The most common fermented milk product is yogurt. The lactobacilli used in the making of many yogurts, however, may not be the same type as found within the common flora of humans as there are many different strains. The following are some of the bacteria used in the diary industry:

- Acidophilus milk is made with Lactobacillus acidophilus.
- Butter is made from pasteurized cream, to which a lactic acid starter has been added. The starter contains, for example, *Streptococcus cremoris* or *S. lactis*, but requires *Lactobacillus diacetylactis* to give it its characteristic flavor and odor.

— Cheese is often made with *Streptococcus* and *Lactobacillus* bacteria. Fermentation lowers the pH, thus helping in the initial coagulation of the milk protein, as well as giving characteristic flavors. In such Swiss cheeses as Emmentaler and Gruyere, the typical flavor is the result of the use of Propionibacterium. Cheese can be classified within two groups — ripened and unripened. Unripened cheeses consist of cottage cheese, cream cheese, and Mozzarella, for example. These are soft cheeses and are made by the lactic acid fermentation of milk. Many different bacteria are used to produce the various cheeses, but *Lactococcus lactis* and *Leuconostoc cremoris* are used most often. Soft cheeses can take one to five months to ripen; hard cheeses, three months to a year or more; and very hard cheeses, like Parmesan, can take twelve to eighteen months. The blue veins found in cheeses, like Stilton and Roquefort, are caused by growth *Penicillium roqueforti*, which is deliberately added now to cheese. Originally, it was found as a natural contaminant of the areas where it was made. The holes in Swiss cheese are the result of *Propionibacterium shermanii*. The surfaces of Camembert and Brie are innoculated with *Penicillium camembertii*, which then develops in a skin on the surface. Limburger is soaked in brine to encourage the growth of *Brevibacterium linens*.

— Kefir includes many different microbes, including yeasts, lactobacilli, lactococci, and leuconostocs. Depending on geographical locations, the precise types of microbes will vary.

— Yogurt usually requires the addition of *Lactobacillus bulgaricus*, *Lactococcus thermophilus*, and/or *Streptococcus thermophilus* to the milk.

Bacteria are not only used for fermentaion in the dairy industry, but for use in other such food production as in the processing of coffee and cocoa, the manufacturing of food additives, and other such processes such as the making of xanthan gum and vinegar. Bacteria, and most viruses do not tolerate acids. This is the reason that vinegar retards the growth of most bacteria. The following is a list of other foods where bacteria and other microbes are necessary for the making of certain foodstuffs.

— Vitamins are the most important non-prescription medications. Most are economically made synthetically, but some are produced biosynthetically, that is, partly from synthetics and partly from bacteria. Such yeast extracts as Vegemite and Marmite are rich in B vitamins.

Microbes may be used to elaborate precursors in the making of Vitamins A, C, and the B family. Using sugarbeet molasses as a growth medium, *Pseudomonas denitrificans* is made to produce Vitamin B12. Members of the genus *Propionibacterium* are also used to make this vitamin. Although many bacteria and fungi produce riboflavin (B2), the fungus *Ashbya gossypii* produces it in huge quantities.

— MSG (monosodium glutamate) is manufactured by *Corynebacterium glutamicum*, a cousin of the diptheria germ.

— Citric acid used to be extracted primarily from lemons, but worldwide demand for it forced the commercial industry to seek other sources. The fungus, *Aspergillus niger*, when grown on sugarbeet molasses, produces huge quantities of citric acid.

— Glutamic acid requires *Corynebacterium glutamicum* for its formation. Biotin is a cofactor essential for lipid synthesis in bacteria. By growing *C. glutamicum* on limited amounts of biotin, it causes the bacterial membrane to leak sufficient quantities of glutamic acid.

— Lysine — The bacterium, *Brevibacterium flavum* is used in the industrial biosynthesis of lysine. Mutants no longer susceptible to feedback inhibition have been isolated to be used industrially to increase the yield of amino acids.

— Baker's yeast (*Saccharomyces cerevisiae*) provides a variety of enzymes that enable carbohydrates to be broken down producing sufficient carbon dioxide to give bread its characteristic texture.

— Beers, etc. — Traditionally, the natural yeasts on grape skins determine the quality of wine produced. These natural yeasts, especially *Saccharomyces cerevisiae* (beer in Spanish is "cervesa") and *Saccharomyces ellipsoideus*, ferment the grapes to make wine. It used to be a risky business leaving it up to nature to decide the quality. Now winemaking has become a regulated science with the use of these yeasts. Beers, lagers, and ales generally rely on the yeast *Saccharomyces cerevisiae*, although lager yeasts will probably always be known as *Saccaromyces carlsbergensis*.

— Sauerkraut-making requires the bacteria *Leuconostoc mesenteroides* and *Lactobacillus brevis* to ferment sugars that provide a variety of such organic products as lactic acid, acetic acid, ethanol, and mannitol. These bacteria are known as 'heterofermentative' bacteria. Later a 'homofermentative' bacteria, *Lactobacillus plantarum* takes over, producing only lactic acid. Later, *Enterococcus faecalis* and *Pediococcus cerevisiae* assume the fermentation process if the salt brine is not what it sould be.

— Sourdough bread requires the help of a yeast, Saccharomyces exiguus, along with lactobacilli, to provide its characteristic texture and flavor.

— Dill pickles are simply fermented cucumbers. Streptococci starts the process of fermentation, but as the pH level falls, *leuconostoc* and *pediococcus* species, as well as *Lactobacillus plantarum* continue the process.

— Olives are edible only after fermentation with *Lactobacillus plantarum* and *Lactobacillus mesenteroides.*

— Coffee and chocolate require *Erwinia dissolvens*, *leuconostoc*, and *lactobacillus* species plus the yeasts of the genus *Saccharomyces* to remove the tough outer coats. The microbes do not affect the taste of coffee but are necessary to confer the characteristic taste to cocoa and chocolate. The bacteria *S. napoli* and *S. eastbourne* often use chocolate as a vector. It is thought that the chocolate provides protection for the bacterium as it passes through the acidic environment of the stomach. This was observed when higher incidents of illness were reported in children. Soy sauce is made from a mixture of soy beans and rice fermented by a variety of bacteria and fungi. These include *Lactobacillus delbrueckii*, *Aspergillus oryzae*, *Aspergillus soyae*, and *Saccharomyces rouxii.*

— Meat products, like salami and bologna sausages, require some fermentation with *Pediuococcus cerevisiae*, *Lactobacillus plantarum* and some members of the genus *Bacillus*. Country cured hams use fungi of the genus *Aspergillus* and the genus *Penicillium* in their fermentation process. Izushi (sushi), a Japanese delicacy made from a mixture of fish, rice, and other vegetables is produced by fermentation with lactobacilli.

Industrial Fermentation

In some cases, the growth of microorganisms in food can be put to good use for the production and preservation of various types of food. Fermentation is arguably the earliest example of biotechnology and refers to the metabolic process by which microbes produce energy in the absence of oxygen and other terminal electron acceptors in the electron transport chain such as fumarate or nitrate.

In ancient times, it was considered as a way to both preserve food and to retain nutritional value. It was probably accidentally discovered in ancient Egypt when dough, made from ground up wheat and rye, was left for a period of time before cooking. In contrast to dough that was immediately cooked, it was observed that the aged dough expanded in size and when cooked produced tastier, lighter bread. The process was not completely reproducible: sometimes the uncooked dough yielded good bread and other times it did not. However if small amounts of good dough was added to the next batch, the bread was again tasty. The Romans went onto improve and perfect this process and popularized this sort of bread throughout the European continent.

The discovery of fermentation in Egypt also led to the first production of wine and alcohol. All these discoveries were largely phenomenological and it would be another 3000 years before the exact cause of fermentation was uncovered. It was Louis Pasteur, again, in 1857 who was able to demonstrate that alcohol can be produced by yeast when grown in particular conditions. This discovery revolutionized the modern food industry: for the first time the agent of fermentation was identified and could be used commercially.

Table 1. Some examples of foods which uses fermentation in their production

Food	*Raw Material*	*Fermentor*
Pickles	Cucumber	*Leuconostoc mesenteroides* *Lactobacillus*
Chocolate	Cacao bean	*Saccharomyces cerevisiae* *Candida rugosa* *Kluyveromyces marxianus*
Bread	Flour	*Saccharomyces cerevisiae*
Coffee	Coffee bean	*Erwinia dissolvens*
Sauerkraut	Cabbage	*Leuconostoc plantarum*
Soy sauce	Soya bean	*Aspergillus oryzae*

Fermentation by bacteria, yeast and mold is key to the production of fermented foods. Fermenting yeast produces the alcohol in beer and wine. In fact, the smell of fresh baked bread and rising dough can be attributed to alcohol produced from yeast.

Fermentation is used to make many ethnic foods such as sauerkraut and miso. Soy sauce is produced by fermenting *Aspergillus ortzae*, a fungus, growing on soy beans. *Erwinia dissolvens*, another type of bacteria, is essential for coffee bean production; it is used to soften and remove the outer husk of beans. Finally, fermentation of milk produces most dairy products.

Without microbes, we would not be able to eat many types of different food that we enjoy today. Table 1 shows example of several foods that are produced through fermentation with specific organisms.

The biochemical process

All organisms need energy to grow. This energy comes from the reduction of adenosine triphosphate (ATP) into adenosine diphosphate (ADP) and results in the release of energy and a phosphate group. In this way ATP serves as a storage molecule of energy which can be used by the cell. But where does the ATP come from? Cells get their ATP from the controlled chemical breakdown of glucose to form two molecules of pyruvate. This process requires two molecules of ATP but results in the release of four molecules or a net gain of two molecules of ATP.

Once pyruvate is formed, it can be processed in several different ways. Mammalian cells usually process pyruvate by putting it into the tricarboxylic or Kreb's cycle. In the presence of oxygen, oxidative phosphorylation produces more ATP from the byproducts of the Kreb's cycle reactions. This is referred to as aerobic respiration. However when oxygen is limiting, other processes must be used in order to deal with pyruvate. This is done through anaerobic respiration or fermentation and involves the breakdown of pyruvate into simpler compounds. Two of the most important fermentation processes which are used on an industrial scale are ethanol or lactic acid fermentation.

Dairy fermentation

Milk is an excellent food source for humans and bacteria alike. It is full of vitamins, fats, minerals, nutrients and carbohydrates. It is rich in the protein casein which gives milk its characteristic white color. The most abundant carbohydrate is the disaccharide lactose, "milk sugar." At room temperature, milk undergoes natural souring caused by lactic acid produced from fermentation of lactose by fermentative lactic acid bacteria. This accumulation of acid (H+ ions) decreases the pH of the milk and cause the casein to coagulate and curdle into curds and whey.

Curds are large, white clumps of casein and other proteins. Whey is the yellow liquid that is left behind after the casein has formed curds. Thus, bacteria obtain nutrients from the milk, inadvertently curdle it and humans use it as the first step in making many dairy products.

The microbes important for dairy product manufacturing can be divided into two groups, primary and secondary microflora. Products undergoing fermentation by only primary microflora are called unripened and those processed by both primary and secondary microflora are called ripened. Primary microflora are fermentative lactic acid bacteria which cause the milk to curdle.

During dairy product production, milk is first pasteurized to kill bacteria that cause unwanted spoilage of the milk and of the downstream milk products. Primary microflora consists of certain kinds of *Lactococcus*, *Lactobacillus* and *Streptococcus* that are intentionally added to pasteurized milk and grown at 30°C or 37°C (temperature depends on the bacteria added). Secondary microflora include several different types of bacteria (*Leuconstoc*, *Lactobacillus*, and *Propionibacterium*), yeasts and molds; they are only used for some types of surface ripened and mold ripened cheeses. The various combinations of microflora determine what milk product you will end up with.

Different unripened milk products are created by using various starting products and bacteria. For buttermilk production, *Lactobacillus bulgaris* is added to skim milk to curdle it. *Leuconostoc* is then added to thicken it. Sour cream is made the same way except cream is used instead of skim milk. During yogurt production, dry milk protein is added to milk to concentrate the milk before addition of actively growing *Streptococci* and *Lactobacilli*. Butter is produced by curdling and slight souring from Streptococci growing in sweet cream. Leuconostoc is then added so it can synthesize diacetyl, a compound that gives butter its characteristic aroma and taste. The milk is then churned to aggregate the fat globules into solid butter. Thus milk type and bacteria will determine the dairy product produced.

Cheese is an important product of fermentative lactic acid bacteria. Particularly in the past, cheese was valued for its long shelf life. Due to its reduced water content, and acidic pH, bacterial growth is severely inhibited. This causes cheese to spoil much more slowly than other milk products. Consequently, the art of cheese production has spread throughout Europe, each country manufacturing many different types of cheeses. Cheese production has three steps: curd formation, curd treatment and curd ripening.

Curd formation can use mare, ewe, cow or goat milk to produce "sour" or "sweet" curd. Sour curd is produced by fermentative lactic acid bacteria as mentioned above. Sweet curd is produced by adding an enzyme called renin instead of bacteria to curdle the milk. The curd is separated from the whey by draining. The curd can be used directly to make unripened cheeses such as ricotta or cottage cheese or can undergo further processing to make a ripened cheese.

Curd treatment consists of condensing and squeezing to form dense, hard curd. It is then molded into the desired shape, salted and mixed with different types of secondary microflora.

Secondary microflora ripen the cheese and will determine the final texture and aroma of each type of cheese. For hard ripened cheeses such as Cheddar, curds are further compressed and the bacteria particular for the cheese is added. The Cheddar is wrapped in wax or plastic to prevent contamination and then incubated to allow the bacteria to do its work. For soft ripened cheeses such as Camembert and Limburger, a microbe, usually mold, is added to the surface of the cheese that produces a protein-digesting enzyme. This enzyme breaks apart the curds and causes the cheese to become creamy and spreadable.

Many cities have long held traditions and nuances for producing a particular cheese i.e. the limestone caves in Roquefort, France which have constant heat and humidity that create unique and delightful cheeses. Microbes can not only be harmful to society but also can be manipulated in a variety of ways for the benefit of society. Particularly in the preservation and production of food, microbes have proven to be useful and essential.

Factors Affecting Microbial Growth in Foods

Many factors affect microbial growth in foods - temperature, oxygen, presence of salts, microbial load and competition, antimicrobial agents, water activity and pH are among the most important. Some of these relate directly to food ingredients. Pickled cucumbers use many of these elements: salt, acid, sugar, fermentation and spices. While pickling dates back 4500 years, many of today's ingredients provide the same effects in a much wider range of foods.

These ingredients work via a number of different and often intertwined mechanisms and, in some cases, the exact effect is difficult to pinpoint. They may react with cell membranes by changing permeability or interfering with movement of chemicals; inactivate enzymes; disrupt genetic functions; or disrupt nutrient utilization or synthesis. Many can work in concert for increased safety. This information can aid a product designer in selecting an ingredient or group of ingredients that will best prevent a microbial time bomb.

Phixing the pH

Most pathogens thrive within a fairly narrow pH range, but they can grow across a fairly wide range that coincides with that of many foods. For example, Staphylococcus aureus has optimal growth between pH 7.0 to 7.5, but will survive in a range of 4.2 to 9.3. A pH from 6.0 to 7.0 is optimum for Escherichia coli, but it will grow from 4.4 to 9.0.

Depending on the microorganism, a low pH can act as a bacteriostat (inhibit microbial growth) or a bacteriocide (kill microorganisms). In most cases the reduced pH acts in concert with other factors - such as increasing the effectiveness of heat treatments or optimizing preservatives - and provides a hurdle to limit growth rather than providing complete safety. That reduced pH doesn't provide complete safety has become painfully evident in the past several years, as acidic beverages such as unpasteurized apple and orange juices have been found to harbor E. coli 0157:H7 and Salmonella.

These acid environments might fall outside the microbial-growth norm, but low-acid foods are ideal for supporting microbial growth. The FDA defines low-acid foods as any that have a finished equilibrium pH greater than 4.6 and a water activity (Aw) greater than 0.85. Without complete sterilization to prevent microbial growth, particularly that of Clostridium botulinum in hermetically sealed packages, foods must have an acid pH. Some, such as tomatoes and oranges, come that way naturally. For those that don't, numerous food-grade acidulants can drop the pH.

The most commonly used acids in the food industry are citric, lactic, malic, acetic (vinegar), fumaric, tartaric, and phosphoric acids. With the exception of phosphoric, these are all organic acids and occur naturally in many foods. Other acidic ingredients, such ascorbic acid, citric acid-containing lemon or benzoic acid-containing cranberry juices, might be added. Additionally, acid-producing fermentations by friendly bacteria might also help achieve a target pH.

Fatty acids have also demonstrated the ability to prevent microbial growth. Lactic and acetic are considered short-chain fatty acids, and they inhibit gram-negative and gram-positive bacteria. Medium-chain saturated fatty acids (C8 to C14), especially lauric acid, restrict the growth of gram-positive microbes. Increased chain lengths, branching and other structural variations typically decrease their potency. Esterification often positively influences the antimicrobial effect; in many cases the monoglyceride is more effective than the fatty acid alone.

The antimicrobial activity of some acids does not come strictly from pH reduction. They also work through the antimicrobial effects of the undissociated ions in solution. Short-chain undissociated acids can pass through cell walls and disrupt cellular metabolism.

Quick chemistry tutorial: Weak acids, a category that includes the organic aids previously mentioned, do not dissociate completely in water. The weaker the acid, the smaller its Ka (acid dissociation constant, a term that describes the extent of the dissociation) and the less it will dissociate. Ka is often expressed as pKa, which equals -log Ka. When the pH of a system approaches the pKa of an organic acid, it allows more unassociated short-chain acids through the walls, so short-chain acids can be more effective at a lower pH.

All acids are not effective against all organisms; certain bacteria might have a built-in defense. For example, Enterobacteriaceae and certain other species can metabolize benzoate. Formulation might also influence how well they work. The pKa of most organic acids limits their optimum pH range, and fats can inhibit the action of some, such as sorbic.

Using acids as antimicrobials is also circumscribed by their effect on the finished product. A pH of 2.3 might stop all growth, but it would also stop most consumers from putting the product in their mouths due to extreme acid taste. The pH can also promote unacceptable reactions, including color changes or precipitation.

Most food acids also complement certain flavors, generally in combinations that occur naturally, such as propionic and cheese, citric and citrus, malic and apples and lactic and meat. "Acetic acid has a very pungent, vinegar-like flavor," says Bert DeVegt, national sales manager, Purac

America, Inc., Lincolnshire, IL. “You can only use a certain level before it gets overpowering. Lactic acid is much milder.”

Certain acids may also exhibit a synergistic effect when used in combination. For example, combining acetic and lactic acids can provide a potent combination, according to DeVegt. When used alone, acetic acid levels of 2.0% or lactic acid of 2.5% are required to inhibit the growth of Lactobacillus plantarum, a spoilage indicator in sauces and dressings. But when using a combination of these two acids, a total level of 1.5% acid is all that is required to achieve the same effect. “If you have a salad dressing with acetic acid and replace 50% of the acetic acid by lactic acid, you have a milder flavor and a more microbial-effective formula,” DeVegt notes. “You can even use a higher level of lactic acid so the entire acid content of the system can be higher and still have a milder flavor profile.” Because lactic acid is a stronger acid than acetic, adding the same weight actually produces a lower pH.

Many acids and their salts can be formulated alone or with other antimicrobials, to make a sanitizing wash. These are typically used for meat products, but may also have some value for other foods such as fresh-cut vegetables. In one study, conducted by researcher Susan Summer, department of food science and technology at Virginia Polytechnic Institute and State University, Blacksburg, fruits and vegetables were contaminated with Salmonella, Shigella and E. coli and then washed with a spray of hydrogen peroxide followed by acetic acid. While the H2O2 alone was fairly effective, the second spray of acetic acid improved the results tenfold.

Lactic and other acids can make effective washes for meat and poultry to keep the microbial count down. “Lactic acid is able to kill bacteria, giving a reduction between log one and log three,” says DeVegt. “Manufacturers use about 2% lactic acid in a water solution as a final wash, before the carcass is chilled. It will kill up to about 99.9% of the bacteria. And if you leave it on the carcass surface, it will also have a bacteriostatic effect and inhibit the growth of bacteria.” This same sort of treatment could also be used for cooked sausages, such as frankfurters, applied as a spray before peeling, he says.

Shake them with Salts

Sodium chloride has been used for its preservative qualities for thousands of years. But other salts - defined as ionic compounds with anions other than OH- or O_2- - also have this property. The mechanics of their antimicrobial action differ somewhat, however.

When some salts dissolve in water, the ions can react with the water, allowing them to act as weak acids. They therefore follow the same behavior as the organic acids. Commonly used salts include the lactates (calcium, sodium and potassium), the benzoates (sodium and calcium), the sorbates (potassium, calcium and sodium) and the propionates (sodium and calcium). In solution, these transform to lactic acid, benzoic acid, sorbic acid and propionic acid, respectively. Formulators can add the acid form but, particularly in the case of sorbates and benzoates, the salt form provides increased solubility over the dry acid.

Salts are useful in meat products, where acids can adversely affect the product. Meats are a pH-neutral product, DeVegt says. “Acids promote protein denaturation. By lowering the pH, you’re also decreasing the yield of the product by lowering the water-holding capacity of the meat protein.” Still, the right pH will increase the effectiveness of the salts, so there may be a balancing act. With sodium lactate (an ingredient widely used to extend shelf life and increase safety of processed meat products), for example, “If you can lower the pH of a meat product by 0.1 to 0.3 units, you end up with a more effective hurdle,” he continues.

Salt (NaCl), on the other hand, works by inducing plasmolysis, or cell dehydration. Because nature subscribes to the theory that all things seek equality, the high salt level on the outside of the cell causes osmosis. Water passes from inside the cell through the membrane into the solution in a vain attempt to achieve equilibrium. Most bacteria, with the exception of extreme halophiles that can live in as much as 30% salt, are more susceptible to elevated salt levels than are yeast and molds. Bacteria that grow best in the absence of NaCl, but can grow at moderate salt concentrations, such as S. aureus, are called halotolerant.

Water activity

Since water is a basic requirement of most life forms, microorganisms require a certain level to survive and grow. In food, the concern is not the total moisture, but the available water as measured by water activity, or Aw. The Aw, or equilibrium relative humidity, measures the vapor pressure caused by the moisture present in a hygroscopic product. This term quantifies a product’s active or unbound moisture in a product. In order to grow, bacteria require a minimum Aw of about 0.90, while yeasts need a minimum of 0.86, and molds, a minimum of 0.70. As always there are exceptions to the rule; for example, some staph species can grow at Aw 0.85, but hopefully none that will kill us.

Any solute can affect the product's Aw, including salt. But unless one is a human halophile, the levels of salt required to reduce it to effective levels would make most foods unpalatable. So, the most common food ingredients to use for Aw reduction are sugars and related compounds - think jellies, fruit fillings, sugary soft bars and other high-sugar, high-solids items. Aw effect is related closely to osmotic pressure, which sugar solutions also exhibit. The ultimate effect of a given weight of solute depends on its molecular weight - the lower the molecular weight, the more moles per gram, and greater the reduction in Aw.

Although all of these ingredients are chemicals, certain ones - through processing, derivation, GRAS non-certification or personal opinion - readily fall into the category of chemical preservatives or antimicrobials. Probably the most famous, and consumer-advocate notorious, are the nitrates and nitrites. The sodium or potassium salts of these compounds are used for curing meat. During curing, nitrates can transform into nitrites. These form nitric oxide, which reacts with myoglobin in the muscle to produce the cured color, nitrosomyoglobin.

Although many claim that these compounds only give the meat a pleasant reddish color, in fact they suppress bacteria, particularly C. botulinum. The exact mechanism is unclear, but may result from reaction compounds that cannot be metabolized under anaerobic conditions.

The problem is that nitrates/nitrites also result in small amounts of compounds that might be toxic to larger organisms, namely humans. When nitrite combines with secondary amines, a reaction promoted by very high cooking temperatures, it forms nitrosamines, which are carcinogenic. For this reason, manufacturers often look to other antimicrobial agents to reduce the level of nitrites required.

Meat has been smoked through the ages to increase shelf life. Several studies conducted at Kansas State University, Manhattan, show that liquid smoke inhibits pathogens such as E. coli 0157:H7, Salmonella, Listeria and Streptococcus. Phenolic compounds in smoke act as bacteriocides.

Sulfur dioxide and sulfates are another example of compounds that, although used for other preservative effects, also might provide protection against bacteria, especially the gram-negative variety. When sulfur compounds go into solution in an acid environment, they can form ions that reportedly prevent microbial growth. These ingredients also have acceptance problems in some quarters - some people suffer severe reactions when exposed to them and their presence must be noted on the product label.

Hydrogen peroxide (H_2O_2) finds use in the dairy and starch industries as a direct additive and as a package sterilizer, particularly in aseptic applications. The FDA restricts its use to a maximum of 0.05% in milk intended for cheese-making, and 0.04% in whey during the preparation of modified whey. In starch, the limit is 0.15%. H202 can also occur as a fermentation byproduct of lactic acid starter cultures. It acts against bacteria - especially anaerobes and facultative anaerobes, yeasts, molds and viruses.

Although not food additives, the sterilizing epoxides ethylene oxide (EO) and propylene oxide destroy all microorganisms, including viruses. These are typically applied as a gas on products such as spices and some nuts. The government has established tolerances for ethylene oxide residues in 40 CFR 180.151 of 50 ppm "when used as a post-harvest fumigant in or on the following raw agricultural commodities: black walnut meats, copra, whole spices." However, EO is regulated by OSHA as a cancer and reproductive hazard.

Another chemical used as an antimicrobial treatment, rather than an ingredient, is ozone (O_3). Ozone is an unstable gas that decomposes into oxygen fairly quickly, especially in water. It acts against many microbial contaminants - including E. coli, Salmonella, Giardia, and Cryptosporidium. Ozone was declared as GRAS for treatment of bottled water in 1982 and in food processing in 1998. Potential food applications include increasing yields

of certain crops, protecting raw products during storage and transit, sanitizing packaging materials, or adding to wash water for foods. Also, Electric Power Research Institute, Palo Alto, CA is investigating ozone control of insect infestation during food storage as an alternative to fumigants currently being phased out, such as methyl bromide, which is scheduled to be banned by the U.S. government in 2001.

Microbe vs. microbe

Other microbes can also help defeat their dangerous brethren by competing for nutrients, lowering pH or producing various metabolites that inhibit other microbial growth. Some of these substances can be purified to add directly to food.

Antibiotics are microbial or synthetic substances that inhibit other microbial species by targeting the bacterial cell wall, membrane and ribosomes. The only antibiotic approved as a direct food additive in this country is natamycin (pimarcin), for use on the surface of cheese products. However, this substance only works against molds and yeasts, not bacteria. It does appear to suppress mycotoxin production, though, so it does contribute to a safer food supply.

The FDA approved a combination of powdered cellulose and natamycin for use in cheeses in response to a petition by St. Louis-based Fiber Sales and Development Corporation that will allow its use in dry applications.

Nisin, a natural antimicrobial, is a polypeptide produced by Lactobacillus lactis. It works over a pH range of 3.5 to 8.0 and has been found to work well against gram-positive bacteria such as Clostridium and Bacillus. Gram-negative bacteria can be made more sensitive to Nisaplin (the brand name for Cultor Food Science's nisin-based preservative) through disruption of their cell walls by exposure to chelating agents, sub-lethal heat treatments, and freeze/thaw cycles, All established uses are against gram-positive bacteria. It is particularly effective against gram-positive, heat-resistant spores and has uses in foods that are pasteurized, but not fully sterilized. Nisaplin is considered to be a natural preservative.

Nisin is allowed in the U.S. in standard-of-identity processed cheese spreads, non-standard-of-identity pasteurized liquid egg products, non-standard-of-identity salad dressings and sauces, and pasteurized soups," Delves-Broughton continues. Typical levels are 2.5 to 12.5 ppm nisin - equivalent to 100 to 500 ppm Nisaplin. "The effect of nisin on a number of different foods has been tested, and some of these uses are approved in different countries throughout the world. For example, nisin has been used to decrease the heat treatment required for canned foods, and appears to decrease botulism toxin production in fish stored under modified-atmosphere packaging. It may also have benefits in some baked goods, meats and alcoholic beverages.

Nisaplin is often used as a factor in hurdle technology,. For example, in processed-cheese preservation, it can be used in combination with lower moisture, low pH, increased salt content, as well as low storage temperatures to produce long shelf life and safe products. Lactic acid bacteria - the same microorganisms used to make fermented dairy and meat products - produce other bacteriocin peptides during fermentation. While they are not currently available or approved as food additives, these metabolites might add to a fermented food's safety. In addition to nisin, these include compounds such as lactococcins, pediocins, lactacins and leucocin A. Many of these act against Listeria.

Food fights

The next wave of antimicrobial agents may be taken from foods themselves. A number of naturally occurring compounds possess some degree of bacteriostatic or bacteriocidal activity. Lysozyme, an enzyme found in the white of hen eggs, exhibits bacteriocidal properties by hydrolyzing certain polysaccharides in bacterial cell walls. Approximately 3.4% of the albumen protein in egg consists of lysozyme. The FDA has approved it for use in cheese to prevent "late-blowing," an undesirable gas formation by Clostridium during cheese ripening. Researchers have also investigated its use as an antimicrobial agent in several other applications, including vegetables, tofu, sausage and seafood. It is believed to act against gram-negative species such as Bacillus and Micrococcus.

"Lactoferrin presents a new opportunity in formulating foods by providing a multifunctional ingredient with antimicrobial and probiotic characteristics," says Loren Ward, Ph.D., research scientist, Avonmore Waterford Ingredients, Monroe, WI. Lactoferrin, an iron-binding glycoprotein, occurs in milk and bodily fluids such as tears, blood and saliva. It can be commercially isolated from cow's milk for use in food products. The ingredient is currently used in infant formulas or functional foods for its positive effect on intestinal flora. Research shows that lactoferrin inhibits E. coli, Salmonella typhimurium, Shigella dysenteria and other bacteria, and may inhibit some viruses also.

Lactoferrin inhibits bacterial growth by at least two different mechanisms: by chelating iron, which makes this element unavailable for microbial growth, or by directly binding to the cell membrane, which alters membrane permeability. "Both antimicrobial mechanisms would be active with an iron-free lactoferrin product such as Bioferrin(tm)1000," notes Ward.

Plants can also be sources of microbial barrier compounds. Spices have been used for centuries as food preservatives. Some are currently being touted for antioxidant capacities, but certain spice compounds also control microbial growth. Most of these are phenolic compounds that are typically not present in concentrations high enough to have an effect when the spices are used normally. Cinnamon contains cinnamic aldehyde, which acts against bacteria and inhibits mold and mycotoxin production. Other spice-based antimycotics include vanillin, citral, citronellol, geraniol and menthol. Eugenol, found in clove oil and allspice, suppresses mold and bacteria such as Salmonella, Vibrio and Staphylococcus. Thymol, contained in thyme, oregano savory and sage, inhibits a wide range of pathogenic bacteria and molds that produce mycotoxins.

Vegetables also might be utilized in the fight against bacteria. They contain a wide range of compounds, many of which are phenolic or flavonoid molecules that have evolved as defensive antimicrobials in the plants themselves. For example, tannins can denature microbial protein. Other plant-based pathogen preventatives include alkaloids, lactones, glucosides, anthocyanins, phytoalexins, caffeine and protein-like compounds, such as zeamatin.

Garlic, onion and similar plants provide allicin, a sulfur compound with considerable antimicrobial activity against both gram-positive and gram-negative organisms. The compound inhibits bacterial enzyme activity. Allicin isn't formed until the plant tissue is disrupted through cutting or crushing, when the plant's allin hydrolyzes. Numerous studies have shown this compound's effectiveness against many food-poisoning organisms, yeasts and molds, although it does not appear to inhibit toxin production of C. botulinum.

Since botulism has been reported after the consumption of both fried onions and garlic in oil, this should come as no surprise, however. One 1973 study by Tynecka and Gos (published in Acta Microbiologica Polonica Series B, a Polish journal) showed a 1:128 dilution of garlic juice inhibits *E. coli* and a 1:64 dilution inactivates the organism.

Another potential natural preservative occurs in high levels in horseradish and mustard oil. Researchers with USDA's Agricultural Research Service and Oklahoma State University (OSU), Stillwater, have looked at allyl isothiocyanate (AITC), a chemical that inhibits the growth of food pathogens such as L. monocytogenes, E. coli and S. aureus. ARS food technologist Henry Fleming, at the Raleigh, NC, Food Science Research Unit, and OSU food chemist Brian Shofran found that mustard oil contains 93% AITC, while horseradish has 60%. Other isothiocyanates turn up in cole vegetables, cabbage, spinach, radishes and related plants.

Continued investigation of these types of compounds and others will become increasingly important in the future. The types of foods we eat, their method of preparation and the seriousness of the effects of microbial contamination all require increased vigilance and improved methods to defeat the microbial marauders that plague the food supply.

Biotechnology and Food Production

Consumers perceive dietetics professionals as reliable sources of food and nutrition information and services. Thus, it is important that dietetics professionals provide consumer education when qualitative, labeling, and regulatory questions arise regarding food materials derived using biotechnology techniques.

The term "biotechnology" has come to mean different things to different people. Because the term is politically charged, it is important to understand the wide range of definitions that are applied to it. The simplest definition of biotechnology is "applied biology." Another definition is "the use of living organisms to make a product or run a process." By this definition, the classic techniques used for plant and animal breeding, fermentation, and enzyme purification would be considered biotechnology. Examples include use of bacteria to make yogurt, cheese, and vinegar and use of plant or animal cross-breeding techniques to produce stock with enhanced qualities.

Some people use the term biotechnology only to refer to the newer tools of genetic engineering developed since 1973. In this context, biotechnology may be defined as "the use of biotechnical methods to modify the genetic material of living cells so they will produce new substances or perform new functions". Examples include recombinant DNA (rDNA) technology, in which a copy of a piece of DNA containing one or a few genes is transferred between organisms or recombined within an organism, and rDNA or gene splicing, which may be likened to cutting a circle of tape, inserting a different piece, and rejoining both ends to the new piece.

Genetic engineering technologies can be used to increase the amount of a protein or metabolite produced by an organism, to allow organisms that did not originally produce that protein or metabolite to do so, or to block production of a protein or metabolite by an organism. rDNA technology allows researchers to move genetic information between unrelated organisms to produce desired products or characteristics or to eliminate undesired ones.

Biotechnology and Agriculture

In the past, plant and animal breeding were used to enhance agriculture by taking advantage of the natural variability of characteristics or inducing mutations (or using natural mutations) in genes. Today, genes from other species can be engineered into plants or animals. For example, antiworm protein from the bacterium *Bacillus thuringiensis* can be engineered into corn or cotton, thus reducing the need for pesticides. Biotechnology has provided the tools for engineering crops resistant to pests or herbicides and animals capable of producing therapeutic proteins. Another example is sheep that produce human antibodies that can boost the immune system. Organisms with genes transferred from one species to another are called transgenic.

The tools of molecular biology which were developed largely for medical applications, are being used for cloning genes into microorganisms to produce bioenergy and bioproducts, as well as for the study and modification of metabolic pathways in microorganisms. These studies require gene sequencing and relating protein function to its structure, much as pharmaceutical properties are related to protein structure in the development of new pharmaceuticals. There is a major difference, however. Outputs of bioenergy (such as ethanol) will be measured in tons per day, whereas outputs of biopharmaceuticals may be measured in kilograms per year.

A yeast has been engineered to produce ethanol from xylose, including the formation of xylulose by xylose isomerase (an enzyme that also isomerizes glucose to fructose). The xylulose then enters an ethanol-producing pathway. Glucose has been fermented to ethanol for millions of years, but when xylose isomerase and other enzymes are cloned into yeast, both glucose and xylose can be fermented. Similarly, a genetically engineered *E. coli* is also capable of fermenting xylose to ethanol.

When combined with a suite of other bioprocessing technologies, pentose fermentation increases the yield of ethanol from plant materials by

50 percent, moving us a step closer to the transformation of renewable, agricultural residues into fuel-grade ethanol. Imagine the potential in the United States alone, where an estimated 20 million tons of residues could produce 1.6 billion gallons of ethanol.

The goal of metabolic-pathway engineering is to understand, and ultimately direct, metabolic pathways in microbial cells to make value-added bioproducts. Some transgenic animals and plants have been engineered to produce therapeutically important proteins (although these are not yet commercial). Agriculture could become a producer of large volumes of therapeutic compounds using small amounts of land. Engineering will play an important role in the extraction, recovery, and purification of these products.

Eighty-one percent of U.S. soybean acreage and 15 percent of corn acreage is already planted with genetically engineered, herbicide-tolerant crops. These crops can survive the application of specific herbicides that would otherwise destroy them, as well as targeted weeds. Insect-resistant cotton and corn contain a gene from the soil bacterium, *Bacillus thuringiensis*, that produces a protein toxic to certain lepidopteran insects (insects that go through a caterpillar stage) and protects the plant over its entire lifetime. Some plants have "stacked" traits (i.e., both insect and herbicide resistance).

In 2003, 27 percent of the cotton planted was insect and herbicide resistant. Corn with stacked traits made up only 4 percent of the corn planted in 2003. Corn with the *Bacillus thuringiensis* trait made up 41 percent in 2003 and appears to be leveling off. Considering that these crops only became available in 1996, their impact has been remarkable. One of the challenges for engineers will be keeping genetically engineered crops separate from other crops, because commercial developers must address consumer resistance to some engineered bioproducts, particularly some types of foods.

Plant Food Applications

Biotechnology techniques are being applied to plants to produce plant materials with improved composition, functional characteristics, or organoleptic properties. Genetic modifications have produced fruits that can ripen on the vine for better taste yet have a longer shelf life through delayed pectin degradation or altered responses to the plant hormone ethylene.

Among the first commercially available whole food products was the Flavr Savr slow-ripening tomato, which US Food and Drug Administration (FDA) approved in May 1994; the gene for polygalacturonase, the enzyme responsible for softening, is turned off in this tomato.

Plants that are resistant to disease, pests, environmental conditions, or selected herbicides or pesticides are also being developed. A variety of squash that is resistant to two plant viruses was approved by FDA in 1994. In 1995, the Environmental Protection Agency (EPA) gave clearance for development of transgenic corn seed, cotton seed, and seed potatoes that contain the genetic material to resist certain insects; FDA approved these biotechnology applications in 1994.

The US Department of Agriculture (USDA) is considering herbicide-resistant soybeans and cotton seed for animal feed. The advantage of such products is that they allow the use of less toxic and more environmentally friendly herbicides and pesticides. Production of heat- and drought-resistant plants could bring agricultural opportunities to regions of the world currently unsuitable for raising food crops.

Plant foods with enhanced processing and/or nutritional characteristics are a likely application of biotechnology. In 1992, Monsanto Company successfully inserted a gene from a bacterium into the Russet Burbank potato. This gene increases the starch content of the transgenic potato. Higher starch content reduces oil absorption during frying, thereby lowering the cost of frying french fries and chips and reducing the oil content in the finished product. In the future, such genetic applications as altering the fatty acid profile in oil seeds and producing wheat with no phenylalanine may be possible.

Applications of biotechnology in plant breeding require more time to develop than pharmaceutical applications because of difficulties associated in working with living tissue. The pace of implementation is also limited by growing seasons. Barriers to applications of biotechnical methods with plants are minor, however, compared with barriers to applications of biotechnical methods in the breeding and production of animal foods.

Animal Food Applications

The first FDA-approved application of biotechnology to animal production was the use of recombinant bovine somatotropin (bST) in dairy cows. bST, a protein hormone found naturally in cows, is necessary for milk production.

As with human insulin, bST can be isolated from cows, inserted into bacterial cells, and produced in large quantities by those cells. The bST can then be recovered, purified, and injected into dairy cows. When the recombinant bST is administered to dairy cows under ideal management conditions, milk production has been shown to increase by 10% to 25%.

After much study and amid some controversy, the FDA approved the use of bST in dairy cows in November 1993. Commercial sales were delayed for 90 days, however, because of a Congressional act. FDA based its rulings on findings that (a) bST is species-specific for cows, (b) bST is a protein that is digested in the intestinal tract of human beings and cows, (c) milk contains bST naturally and supplementation does not increase the amount of bST to levels outside the normal ranges, (d) bST supplementation does not change milk composition, and (e) bST has not been found to cause growth-promoting activity in a variety of species. Although results of a 1-year follow-up study of cow health issues have produced no new animal health concerns, the use of recombinant bST in dairy cows remains a political and social issue.

Porcine somatotropin (pST), a hormone similar to bST but active in hogs, is a "repartitioning agent" in that it redirects dietary energy away from fat deposition in the direction of lean muscle tissue production. Use of pST increases pork carcass leanness and reduces fattiness. Although pST has not yet been approved for use in the United States, it has been approved in European countries.

Other uses of biotechnology in animal production include development of vaccines to protect animals from disease (eg, swine pseudorabies), production of several calves from one embryo (cloning), artificial insemination, improvement of growth rate and/or feed efficiency, and rapid disease detection. In addition, "transgenic animals" or animals with altered genetic make-ups could be produced. The pace of implementation of biotechnology in animal foods will likely be limited, however, by difficulties inherent in working with living tissues and life cycles and by consumers' social and ethical concerns regarding animal applications of biotechnology.

Food Ingredient and Processing Applications

In addition to the genetic manipulation of whole plant foods and animals, microorganisms can be designed to improve the efficiency of fermentations and other primarily enzymatic processes and to produce natural food ingredients. Biotechnical methods can produce food materials with improved nutritional value, functional characteristics, shelf stability, and/or sensory characteristics; more efficient processing techniques; more sensitive analytic techniques for quality control and food safety; and bioremediation techniques that convert by-products to fuel, chemicals, or usable materials.

Microbes have been genetically engineered to produce amino acids for the synthesis of aspartame. In addition, plant cells grown in fermenters can produce flavors such as vanilla, reducing the need for extracting the compounds from vanilla beans. Food processing has benefited from biotechnically produced chymosin (rennet), which is used in cheese manufacture; alpha-amylase, which is used in production of high-fructose corn syrup and dry beer; and lactase, which is added to milk to reduce the lactose content for persons with lactose intolerance.

The demand for new and improved commercial products increasingly will be met through bioprocessing, a type of advanced manufacturing that involves chemical, physical, and biological processes employed by living organisms or their cellular components. Bioprocessing enables the translation of research discoveries into commercial products with unique and highly desirable characteristics and offers new production opportunities for a wide range of items, including

— energy sources, including alternative fuels such as ethanol, methane, and diesel hydrocarbons;
— commodity chemicals, such as enzymes, organic acids, and solvents;
— polymers for films, coatings, and other specialty uses; foods, such as glucose and glucose-fructose syrups; and
— improved high-value products, such as naturally occurring non-toxic pesticides, flavoring agents, and pharmaceuticals.

Bioprocessing offers a level of specificity, predictability, and productivity that otherwise would not exist in the manufacture of these products. Moreover, when the raw materials contain certain molecules with complex structures, bioprocesses enable the synthesis of products that cannot be made by any other means. Combined, these capabilities provide for new process designs that are cost effective, energy efficient, product specific, and environmentally benign.

While bioprocessing already is employed in industry, research is needed to define the parameters of the related processes precisely in order to expand use of this approach and derive the maximum possible benefits. Governments play a critical role in this area because individual manufacturers are unlikely to invest in time-consuming generic research for the benefit of entire industries.

Process design involves the integration and scale-up of a number of operations, including upstream and downstream processing, and process monitoring, optimization, and control. One objective for process design is to quantify process variables and product yields in order to obtain a cost estimate for producing a material of biological origin, and to evaluate alternative designs that minimize the cost of producing the product. An optimal process can be one that minimizes the cost of producing a material, maximizes the profit, or assures that an operation meets special safety or environmental requirements.

Upstream processing

Upstream processing encompasses any technology that leads to the synthesis of a product as well as the fundamental science and engineering needed to understand product formation. Specific areas of opportunity for research include biocatalysis, metabolic engineering, biomass conversion, bioreactor design and cell culturing techniques, and transgenic animals.

Biocatalysis

Nature's catalysts — enzymes — function highly selectively to accelerate the rate of chemical reactions in biological systems. Enzymes also can limit reactivity toward a particular substrate and limit chemical conversion to a single desired product. For many substrates, enzyme-catalyzed reactions exhibit rates that are many orders of magnitude faster than those of uncatalyzed reactions. In chemical synthesis and other processes of industrial importance, the ability to mimic the selectivity and rate-enhancing properties of enzymes would be of substantial benefit in reducing economic and environmental costs. In some cases, this approach might provide a commercial means of synthesizing a material where only much less desirable methods were available previously.

At present, about 50 enzymes are used in industry. Most of these industrial uses involve breaking down large molecules into simpler ones. Examples include the proteases in laundry detergents, and the hydrolases used in starch processing. The commercial benefits of enzymology would be enhanced if technologists could expand capabilities for creating complex molecules from simpler ones or transforming existing chemical structures into more active compounds.

Production of sweetness- enhanced corn syrup by the enzyme glucose isomerase is one noteworthy example of how a chemical transformation can be carried out on the industrial scale. Synthesis routes that can distinguish between right- and left-handed chemical structures are being pursued for pharmaceutical, food, and agricultural applications. Many industrial firms are exploring the unique capabilities of enzymes to carry out such chemical transformations.

Government sponsored research can facilitate the development and application of new enzymes of industrial importance. Many research efforts are under way, and solid progress has been made using protein engineering, chemical modification, and recombinant monoclonal antibody technology. More recently, recombinant DNA technology has been used to generate catalysts that exhibit the useful characteristics of naturally occurring enzymes but have been designed either to catalyze chemical reactions not found in nature or to alter biological reactions to broaden their specificity and expand their potential for practical application.

Progress in this field will accelerate with an improved understanding of relationships among protein structure, function, and energetics. More efficient methods are being developed to obtain protein structure using X-ray crystallography, nuclear magnetic resonance (NMR), and neutron scattering techniques, coupled with robust algorithms to treat the data and display the results using computer graphics. The use of computer graphics to visualize the dynamics of proteins also will enhance understanding of the mechanisms of enzyme action. Progress in obtaining structural data in real time ultimately will help modelers develop algorithms to predict the dynamics of enzymes. Research also should be supported on the thermodynamics and kinetics of prototypical enzyme-catalyzed reactions of interest to the bulk and commodity chemical markets; examples include the reactions involved in corn processing and the action of laundry detergents.

Metabolic engineering

Metabolic engineering is the use of recombinant DNA technology to enhance the activities of a cell by manipulating its metabolic pathways. To exploit this approach, scientists must develop an improved understanding of cell metabolism. Experimental and mathematical techniques are needed for quantifying the effects of altered uses of raw materials by cells. Techniques such as electrophoresis can be used to distinguish and quantify cellular proteins produced in response to altered metabolism. In addition, non-invasive techniques such as near-infrared spectroscopy and NMR can quantify intracellular concentrations of raw materials. Also of interest is research on thermodynamics and kinetics of prototypical microbial and microalgal systems.

The first applications of metabolic engineering have incorporated genes for pathway enzymes and transport proteins. In this manner, for example, the bacterium *E. coli* has been induced to produce ethanol. But amplified expression of non-native proteins can affect host cell metabolism — sometimes interfering with the engineering objective — because raw material resources are reallocated. The effects may include a slowing of cell growth and possibly cell death. Often, the expression of small amounts of non-native proteins increases the level of stress proteins.

This problem is ubiquitous, occurring in simple organisms such as *E. coli* and in complex cells, such as yeast cells, human cells, the chinese hamster ovary cell and an insect cell line that is a possible source of proteins for making AIDS vaccines. Government's support for research in this area can promote efforts to overcome or mitigate these undesirable cellular responses.

Biomass conversion

Biomass includes organic polymeric material — such as lignin, starches, celluloses, and oils — produced by biological processes. Plants and algae are the main sources of biomass, generating many billions of tons annually through photosynthesis. Other sources of biomass derived from human activities include food processing wastes, waste paper, and municipal solid wastes.

Governments should expand thier support for research that will facilitate development of commercial products from this abundant raw material. Biomass is a potential source of energy resources such as methane and ethanol, commodity chemicals, animal feed, and specialty products (e.g., flavors, fragrances, pigments). Biomass is an attractive alternative to petroleum- based sources because it not only is a low-cost substrate for production, but also is environmentally friendly; it is renewable and may reduce emissions during processing and improve the biodegradability of end products.

Biotechnology is especially relevant to the commercial use of biomass because it employs natural processes uniquely suited for creating specific products. An example of such a process is fermentation, which converts glucose sugars derived from biomass into commodity chemicals. The components of plant biomass, and products that can be derived from them, include

— cellulose (a mostly crystalline polymer of glucose molecules), which can be hydrolyzed (i.e., broken down) to form glucose, which in turn can be bioconverted into a variety of products, such as ethanol;

— hemicellulose (another heterogeneous, amorphous sugar polymer), which can be hydrolyzed to make sugars such as xylose and glucose;

— lignin (a complex aromatic polymer), which can serve as a chemical feedstock or be burned to provide heat and electricity;

— starch, used by the corn wet milling industry and fuel alcohol industry as the basis for large, commercial-scale bioprocesses; and

— oils. Production of diesel fuel from oil seeds is probably the most well known industrial application of a "renewable" oil; the hydroxy fatty acids from oil seeds can be used in a wide range of products, including cosmetics, waxes, nylons, plastics, and coatings.

  Fuels and chemicals can be produced not only from oil seed crops (e.g., soybeans, canola, and crambe), but also from unicellular algae and waste fats from animals. The substantial fractions of oils in these resources can be transformed through a chemical process that changes the large molecules in fats and oils into smaller molecules.

Bioreactor design and cell culturing techniques

If the great variety of potentially commercial bioprocesses are to be developed and applied, then bioreactors must be designed in which the environment can be controlled precisely to maximize process efficiency. Government's support for research in this area can improve the efficiency of bioprocessing and thereby improve prospects for its wide application.

The design of a bioreactor requires a basic understanding of both chemical reactor design and cell biology. First, designers must understand the effects of reaction rates and stoichiometry, mass transfer, heat transfer, and turbulence and mixing on product distribution, reactor productivity and size, and operational characteristics. These phenomena need to be expressed in accurate but tractable models that can be used for design and optimization calculations.

In order to develop effective cell culturing techniques, designers also must have basic knowledge of cellular functions and protein chemistry. They should understand the molecular, genetic, and metabolic processes involved in the growth of cells and the expression of cellular products; and structure/ function relationships in the use of proteins for biochemical conversions.

Transgenic animals

Transgenic animals can be employed either as biofactories for the production of commercial products or as living models for the study of human diseases and evaluation of pharmaceuticals. The use of transgenic animals for these purposes can be more economical or, in the case of human disease and drug models, more realistic than are conventional alternatives. Government supported research can improve the efficiency of the processes involved and identify new opportunities.

Transgenic animals are being developed for a wide variety of applications. Transgenic cows are used for production of lactoferrin, a naturally occurring milk protein used in products such as baby formula. Several companies plan to market pharmaceuticals from the milk of sheep and goats and the blood of pigs. Anti-clotting factors used to treat cardiovascular diseases have been produced in rabbits, goats, and mice.

Research also is under way to develop transgenic animals — mice in particular — that carry genes associated with human diseases. Such animals might enable researchers to identify specific disease genes or produce human antibodies.

Downstream processing

Downstream processing includes the cost-effective separation and purification of bioproducts, as well as biorefining. Often, downstream processing is the most expensive phase of producing a substance of biological origin, especially for products with stringent regulatory requirements. A key reason for this high cost is the complex and dilute nature of the aqueous solutions in which bioproducts generally are produced; an inverse relationship has been demonstrated between the price of biological products and the strength of the concentrations from which they must be isolated. This is the case for bulk and commodity chemicals, including ethanol from biological sources.

The high cost of downstream processing means there is significant potential for savings from improved processes. That factor, coupled with the broad applicability of downstream processing research, makes this area a prime target for investment. For regulated bioproducts, such as pharmaceuticals, removal of trace impurities is the expensive step in the purification.

Separation and purification

The separation and purification of materials produced in a bioreactor is a critical part of a manufacturing operation. The biological products involved range from high-value-added substances used as pharmaceutical agents (e.g., insulin) to lower-cost products including commodity chemicals (e.g., ethanol).

High-value bioproducts are usually fragile molecules, such as proteins or peptides, that require highly specialized and mild processing conditions and may need to be separated from a complex mixture of molecules, including cell debris. This combination of factors makes separation difficult. At present, most separation schemes are scaled-up laboratory procedures; research is needed to improve their performance.

Biological methods hold promise for improving separations. For example, a cell could be modified genetically to suppress production of undesirable by-products and alter the cell wall so it would be permeable only to the desired product. It is also possible to genetically engineer a product so that it passes through the cell wall, has an affinity for certain separation matrices, or is joined to another molecule with desirable separation characteristics. In addition to improving separations, biological approaches also reduce energy expenses, a major part of the cost of separating biological products.

Biological processes one day may offer economical alternatives to current, petrochemically based methods for manufacturing organic acids and alcohols. However, before these bioprocesses can become commercially viable, nontraditional, lower-cost separation methods need to be developed. Research is under way to develop extracting solvents, resins and sorbents that are more selective and have a higher capacity than do current materials. Reversible extraction systems are needed that respond to changes in temperature, pressure, or acidity. Combinations of conventional separation methods and biological methods are being explored to reduce product inhibition, which often occurs in fermentations that produce alcohols and solvents.

In addition, mathematical models of separation steps need to be developed to help reconcile regulatory requirements with basic process conditions during early stages of process development, and to meet the demands of a competitive business environment. Modeling enables scale-up considerations to be estimated very quickly, a capability needed in order to commercialize a bioprocess in an industry where being the "first to market" is a critical element of success.

Biorefining

Biorefining involves the use of microbes in mineral processing systems. Biorefining is environmentally friendly and in some cases enables the recovery of minerals and use of resources that otherwise would not be possible. For example, a significant amount of copper is biorefined from slag heaps. Research now under way or planned addresses the use of microorganisms to bioleach oxide and sulfide ores, and to concentrate metals.

Current research on bioleaching of oxide and sulfide ores addresses treatment of manganese, nickel, cobalt, and precious metal ores. The objective is to identify metabolic pathways in the microorganisms responsible for metal solubilization, and to improve their survival rates and stability. Increased understanding of metabolic pathways will open the door to the manipulation of parameters such as kinetics and metal selectivity, with the aim of enhancing mineral recovery.

An intriguing potential use of biotechnology is *in situ* bioleaching of ore deposits or waste piles. Research is needed to develop a mechanistic understanding of the bioleaching process and to identify environmental and process factors affecting biosystem performance. In addition, bioprocess monitoring schemes and cell-free leaching systems should be developed. Cell-free or cell component systems, in which cellular components rather than live organisms are used for a specific reaction, eliminate the need to provide an environment conducive to survival of a microorganism. Such approaches also tend to enhance control over the reaction, due to the specificity of the cell component.

Biosorption and metal recovery from dilute aqueous solutions is an emerging field of interest, from both a resource conservation standpoint and an environmental remediation standpoint. Microbes, algae, cell wall material, proteins, and other types of biomass have been investigated for use in this application, but economical methods for selective metal recovery have yet to be developed. Governments should support research in this area. Inexpensive substrates for protein immobilization should be developed, and biomolecules with high metal-binding capacity identified.

Use of microorganisms to remove fine particles from aqueous process streams is also of interest. Laboratory-scale studies have shown that certain microorganisms can cause flocculation of fine mineral suspensions, while others can function as flotation collectors or depressants. Extensive research will be needed before the use of extracellular colloidal molecules (such as polysaccharides and exopolymers released by live biomass) can become a viable alternative to current chemical-based systems.

Process monitoring and control

Process monitoring and control requires knowledge of the current state of every step in a bioprocess and the ability to control the process for optimal results. The most efficient use of upstream and downstream processing depends on knowledge of the state of the system and on control algorithms that can optimize the process and maintain it. This is another area where government support for generic research can have significant payoffs for multiple industries.

Control and optimization is challenging due to several characteristics of bioprocesses, including their high degree of nonlinearity (meaning that nonlinear differential equations are required for mathematical modeling), especially in batch and fed-batch bioreactors; and their potential for instability when they involve high-yield mutant or recombinant organisms. There are technical obstacles as well, including the unavailability of reliable on-line real-time sensors and realistic models that capture the complexities of biological systems by identifying and quantifying rate-controlling steps in reactions.

Optimization and control methods should be robust, adaptive, and suited to nonlinear processes. Important components of these systems include biosensors and bioelectronics/bionetworks. Because an optimal process often is operating at the limits of process variables, process control can be very difficult; consequently, design methods should consider simultaneously both the economics and the controllability of the process. The affordability of some production processes, particularly for lower-value products, depends on the development of highly accurate sensors and models and the capability to reproduce bioreactor operation precisely.

Biosensors

Sensors are invaluable in the design and operation of automated and environmentally benign manufacturing processes, and in detection, monitoring, and control of food additives, food safety factors, and bioremediation technologies. Biosensors are attractive because they can harness the inherent specificity of many biochemical processes to allow for accurate and highly specific detection of organic and inorganic species in real time. Biotechnology can be used to design, construct, probe, and alter sensitive and highly specific biosensors and incorporate them into transducing and reporting systems.

A biosensor encompasses three major components: a biological component, an interface, and a transducing element. The biological component (e.g., enzyme, immunoprotein, nucleic acids, whole cell) interacts specifically with the substance to be detected. The interface component (e.g., polymeric thick or thin film, chemically modified surface) links the biological component with the transducer. The transducer converts the biochemical interaction into a quantifiable electrical or optical signal.

An example of a biosensor is the plasmon sensor, which consists of a monolayer of immunoproteins immobilized on the surface of a thin film of gold. The analyte — whatever compound binds to the protein employed — attaches to the immunoprotein and changes the properties of surface electromagnetic waves that can be excited in the gold film. This change, which signifies the presence of the analyte, can be detected optically.

A vast store of fundamental information is available on the biological components used in biosensors. The key challenge now is to improve the manufacturing process so that reproducible, dependable, low-cost biosensors can be mass produced. Government support can hasten progress in this area and foster the use of biosensors in a broad range of applications.

There are a number of technical obstacles. Poor adhesion of polymer films to the transducer, and hence poor reliability, is a major problem. The film must be compatible with the biological component and also adhere to the transducer surface. Another problem is that insufficient knowledge of methods and effects of immobilizing biological components in/on polymeric films has led to wide variation in sensor response, complicating the manufacture of commercial biosensors. Non-specific binding must be minimized, because the specificity of analyte detection is degraded when other molecules compete for binding sites.

There are also operational difficulties that preclude the use of biosensors in bioprocessing environments. For example, many sterilization techniques either modify or denature biological molecules. There are also problems with miniaturization, sensitivity, dynamic range, and noise reduction related to specific biosensor fabrication strategies. Finally, there is the question of quality control — determining quickly and nondestructively whether the biosensor is operating with prescribed precision and accuracy.

Biomaterials processing

This section addresses the processing of one particularly promising category of products — biomaterials. The tools of biotechnology can be employed to endow materials with properties not achievable using more conventional means. This research area deserves steady government support because of the important applications for biomaterials, which include biological substitutes for human tissues.

Biomolecular materials

Due to their diversity, versatility, and unusual combinations of properties, biomolecular materials offer promise for application in virtually all sectors of the economy, including defense, energy, agriculture, health, and environmental technology. Examples of biomolecular materials include silk obtained from spiders, and the ceramics in sea shells.

In addition to their direct use as natural cellular products or modified derivatives, biomolecular materials serve another very important purpose by demonstrating how nature has optimized their physical properties. Research is needed to clarify how higher- order structure is achieved and how it serves to determine macromolecular function in such a variety of forms. With continued advances in modern biology, molecular genetics, and protein engineering, and with rapid improvements in physicochemical characterization, novel biomolecular materials can be designed and tailor-made to meet specific needs. This, in turn, would expand the possibilities for practical applications of these materials.

Standard chemical synthesis has inherent limitations, including the production of compounds with unwanted impurities and by- products. By contrast, biomolecular synthesis — manufacturing methods based on biological processes — allows precise control, thereby reducing levels of impurities and by-products. For instance, when a cell produces peptide polymers, control of the amino acid sequence is assured by the fidelity of RNA and DNA replicative mechanisms. Thanks to recombinant DNA technology and continuing advances in molecular biology, that same control over uniformity of composition, length, and sequence now is available to the scientist seeking to synthesize and express natural or tailor-made genes for peptide polymers.

By inserting appropriate DNA sequences into microorganisms, algae, and higher plants, investigators have learned how to induce the cells to synthesize polymers (such as polylysine) that they do not ordinarily produce, in quantities sufficient for further study and eventual application. Research is needed to define the classes and specific types of polymers that can be produced in this manner. Government support could play a key role in advancing research in this area.

Tissue engineering

Tissue engineering — a term coined in 1987 referring to the development of biological substitutes to restore, maintain, or improve human tissue function — employs the tools of biotechnology and materials science as well as engineering concepts to explore structure-function relationships in mammalian tissues. This emerging technology could provide for substantial savings in health care costs and major improvements in the quality and length of life for patients with tissue loss or organ failure.

Advances in the study of tissue growth and regeneration, at both the cellular and tissue levels, set the stage for practical application of tissue engineering. The culturing of cells in two-dimensional monolayers enabled the study of cellular processes and opened the door to genetic manipulation. Scientists and engineers have begun to view cell culturing as a three-dimensional process, in which external forces on the cells not only may influence cellular products, but also may reawaken the cellular differentiation process. In order to develop living tissue equivalents, it will be important to understand how the cellular environment affects the differentiation process as well as interactions between the engineered tissue and the host.

The first success with differentiated cells came with engineered human skin, now in clinical trials. Scientists also are beginning to explore the potential to grow many tissues in culture. Using stromal cells from human tissue, researchers are developing blood vessels, bone, cartilage, nerve, oral mucosa, bone marrow, liver, and pancreatic cells. Government support can hasten progress in development of these materials. Encapsulated cell therapy is an example of a technique under development by industry that employs biomaterials in the treatment of certain serious, chronic diseases. The goal of this approach is to replace cells within the body that have been destroyed by disease in order to augment circulating or local levels of the deficient molecules. Targets for replacement include insulin-producing cells in diabetics and dopamine- secreting cells in patients with Parkinson's disease.

An encapsulated cell implant consists of cells that secrete the desired hormones, enzymes, or neurotransmitters, enclosed within a polymer capsule and implanted into a specific site within the host. Animal studies have shown that the functional activity of secreting cells can be maintained in vivo. The capsule wall is designed to allow passage of small molecules but prevents or retards the passage of large molecules, such as elements of the immune system. Studies suggest that the transplanted cells are protected from destruction and perhaps even recognition by the host's immune system, allowing the use of unmatched or even genetically altered tissue without systemic immuno-suppression. Thus, the use of an encapsulating membrane may overcome two of the difficulties that prevent widespread tissue transplantation into humans: the limited supply of donor human tissue, and the toxic effects of immunosuppressive drugs required to prevent rejection of unencapsulated transplants. This is an important research area where expanded government investment could yield significant dividends.

Food Safety Applications

Biotechnology offers effective techniques to address consumer concerns about microbial contamination of foods. Biotechnical methods may be used to decrease the time necessary to detect foodborne pathogens, toxins, and chemical contaminants and to increase detection sensitivity. Enzymes, antibodies, and microorganisms produced using rDNA techniques are being used to monitor food production and processing systems for quality control.

Microbial probes, biosensors based on adenosine triphosphate (ATP) content, are being used experimentally as indicators of bacterial contamination. Biosensors to detect animal disease, alterations in product quality, or temperature abuse are under investigation. These developments offer the potential of lowering the cost and improving the safety of the food supply in a timely manner.

Waste Management Applications

Waste management, or bioremediation, is an area of increasing interest to consumers. In response, dietetics and food professionals are initiating efforts to control the amount of waste generated in foodservice operations. Through application of biotechnical methods, enzyme bioreactors are being developed that will pretreat some disposable serviceware or food waste components and allow their removal through the sewage system rather than through solid waste disposal mechanisms or will allow their conversion to biofuel for operating generators.

Microbes can be induced to produce enzymes needed to convert biodegradable materials into the building blocks for new polymers. Waste streams can be controlled to convert by-products to biofuel (wheat straw to glucose to ethanol), specialty chemicals (sugar or fat substitutes), or feedstocks and other useful materials (packaging materials or coatings).

Issues of Biotechnology in the Food System

Public health/food safety issues

With any new technology, safety issues must be addressed. Potential public health hazards posed by introduction of a genetically engineered product into the food supply include an increase in toxins naturally occurring in the parent line, inclusion of a new and potentially allergenic protein derived from the originator of the genetic material, introduction of "unnaturally occurring" hormones into the food supply, and the possibility that bacterial resistance from genetically engineered organisms could transfer to pathogenic strains of bacteria.

Biotechnology applications in food and agriculture are the subject of extensive regulatory review to protect against potential negative effects on food safety and the environment. Federal agencies involved in biotechnology regulation include USDA, which evaluates whole foods and production processes; FDA, which evaluates whole foods, food ingredients, and food additives; and EPA, which evaluates production process.

The FDA currently evaluates each application of biotechnology to animal food products on a case-by-case basis. In contrast, FDA has determined that plant foods produced through biotechnology present no inherent risk and, therefore, should be regulated as any other food entering the marketplace. Under the jurisdiction of the federal Food, Drug, and Cosmetic Act, FDA uses characteristics of the food, not the processes used in its production, as the basis for regulating food products derived through biotechnology. FDA offers a decision-tree approach for companies developing plant food using biotechnology. These decision trees represent a series of testing procedures that enables food processors to anticipate safety concerns and consult FDA as necessary for regulatory review of products under development.

The assessment focuses on toxicants characteristic of the host and donor species; the potential that food allergens will be transferred from one food source to another; the concentration and bioavailability of important nutrients for which a food crop is consumed; the safety and nutritional value of newly introduced proteins; and the identity, composition, and nutritional value of modified carbohydrates or fats and oils. To ensure safety, a variety of toxicologic and product safety data must accompany the application for approval of any food product or ingredient produced via genetic engineering.

Labeling

Labeling food from genetically modified plants and animals has become an important issue. Some consumers and consumer groups believe they have a right to know whether genetic engineering was used to produce a food, some want to be able to choose food on the basis of how it is produced, and some believe labels are needed to notify consumers of potential allergens. Others believe labeling is not necessary if foods are essentially equivalent in composition.

Food labels are regulated by FDA and, in some cases, by USDA. Regulatory agencies are concerned with ensuring that food labels are both true and not misleading. As a result, label information may be compulsory, permitted, or prohibited. In all the assessment considerations, producers of whole foods or food components produced biotechnically must provide evidence that no safety issues are raised. New products must comply with existing portions of the Food, Drug, and Cosmetic Act. For example, if producers introduce a potential allergen into a food, a compulsory label declaration must be made so that allergic consumers can avoid the product. Failure to do so would be regarded as misleading labeling and would result in enforcement action against the producer.

Not all foods developed using rDNA techniques are required to bear a special label. As stated in the Federal Register:

> "FDA believes that the new techniques are extensions at the molecular level of traditional methods and will be used to achieve the same goals as pursued with traditional plant breeding. The agency is not aware of any information showing that food derived by these new methods differ from other foods in any meaningful or uniform way, or that, as a class, foods developed by the new techniques present any different or greater safety concern than foods developed by traditional plant breeding. For this reason, the agency does not believe that the method of development of a new plant variety is normally material information... would not usually be required to be disclosed in labeling for the food."

On a case-by-case basis, some types of label declarations are permitted. For example, a food processor seeking to sell cheese to vegetarians would be permitted to include the statement on the cheese: "Made using microbial enzymes; no animal rennet used".

Misleading label statements are prohibited, even if they are true. For example, FDA guidelines do not allow a label statement such as "the milk was derived from cows not injected with bST" unless an accompanying statement makes it clear that there is no substantial difference between milk from bST-treated cows and milk from untreated cows. The FDA's rationale for requiring the qualifying statement is that a "no bST" declaration should not be construed to imply that milk from bST-supplemented cows is less wholesome.

Although FDA evaluates food products and ingredients produced using biotechnology on a case-by-case basis, the labeling regulations bear out the historical whole-food approach; if the whole food is not materially different from its traditional counterpart, mandatory labeling designating it as a product of biotechnology is not required and is, in fact, misleading unless accompanied by a statement clarifying that there is no difference in healthfulness between the two products. Whereas FDA's approach to labeling biotechnically produced foods is based on sound science, it is not without controversy. Some groups believe it does not adequately provide for consumer right-to-know and social issues; others are content with the present arrangements.

Social and consumer issues

If biotechnology is to be used to ensure a safe, abundant, and affordable food supply, it must be accepted by the public. Increasingly, public interest groups are questioning whether technological change is good or needed, particularly as it affects food safety, the environment, animal rights, and the changing structure of agriculture.

Recent surveys regarding consumer attitudes about biotechnology have shown that consumers are not well informed about biotechnology, but are interested in it and are cautiously optimistic about its use in food production and processing. In a national telephone survey conducted by Hoban and Kendall, most participants expressed positive attitudes about the general use of biotechnology in agriculture and food production.

Certain applications were more acceptable than others, however, depending on the perceived value and potential effects. For example, using biotechnology to change plants was considered much more acceptable than using it to change animals. Transgenic applications of biotechnology, such as the insertion of animal genes into plants, were unacceptable to many participants.

Environmental concerns were important to most people and many considered ethical issues important as well. Bruhn reported that consumer concerns about biotechnology related to perceived unpredictability, risks to the environment, alterations in the ecosystems, and moral and social questions. Bruhn and Hoban and Kendall stressed the importance of education that uses practical explanations and familiar contexts.

Implications

Consumers perceive dietetics professionals as reliable providers of food and nutrition information and services. Hoban and Kendall found that consumers would be most likely to trust information about biotechnology that was provided by a dietitian/nutritionist. A survey of dietetics professionals, however, indicated that most respondents were not comfortable answering clients' questions regarding milk or dairy products from bST-treated cows.

Nearly all of those surveyed wanted more information on biotechnology and food. If dietetics professionals are to maintain the trust of their clients, it is imperative that they understand the terminology and basic science behind applications of biotechnology, along with the complexities of issues surrounding the use of biotechnical methods. A critical issue is food safety.

Issues of a social, ethical, economic, and environmental nature are also of concern. Only when dietetics professionals understand and appreciate the complexities of such issues can they help consumers make informed choices. Improved knowledge will permit consumers to focus on substantive issues and evaluate the validity of these new technologies effectively.

Future Applications of Microbial Biotechnology

Microbial biotechnology will play a significant role in solving many of the problems that are facing the world. Ten major problems that are challenging the world have been discussed recently at a meeting in Copenhagen, Denmark and among the Copenhagen Consensus are three problems that are relevant to microbiologists (Table 1). These challenges are:

1. Communicable Diseases,
2. Malnutrition and Hunger, and
3. Sanitation and Water.

In these three areas of major concern, microbiologists will be able to contribute significantly in the future and play a role by application of current and future microbiological technologies. What types of technologies will be applied in the future? Among several possibilities, Table 2 lists a few approaches.

Table 1. The 10 Challenges of the Copenhagen Consensus

1.	Armed Conflict	2.	Climate Change
3.	Communicable Diseases	4.	Education
5.	Financial Stability	6.	Governance and Corruption
7.	Malnutrition and Hunger	8.	Population and Migration
9.	Sanitation and Water	10	Subsidies and Trade Barriers

Nanatechnology applied to microbiological systems will allow the rapid and sensitive identification of DNA sequences that can establish the types of microorganisms that are present in foods and water and that will lead to safety in our food and water supplies. As part of the feature of nanotechnology, rapid identification of dangerous microorganisms in terrorism activities can also be achieved. The use of virochips and bacteriochips containing known sequences of dangerous and potentially danger disease causing organisms will allow rapid identification of causes of outbreaks of disease, e.g. SARS or avian flu. This technology has already paid dividends in the rapid detection of the causative agent of SARS and with further development and knowledge of nucleic acid sequences, the virochips should be even more useful for rapid detection of disease causing viruses.

Table 2. Examples of future microbiological technologies and areas of study

Nanotechnology:	Food safety tests and various diagnostics
	Detection of bioterrorism weapons
Virochips and Bacteriochips:	Diagnostics for rapid detection of disease bearing microorganisms
	Detection of bioterrorism weapons in air and water
Marine Microbiology:	Detection of unknown microorganisms and useful products
	Culturing of "unculturable microorganisms"
Use of renewable biomass for industrial purposes and for energy:	Conversion of industrial microorganisms to cellulose utilizers
	Intercellular complementation - cellulosomes produced by mixed cultures

Marine microbiology appears to be ripe for future studies, since the recent survey of nucleic acid sequences by Venter and his colleagues has revealed he remarkable diversity of microorganisms in the marine world. In approximately 1500 liters of sea water, they found 1800 species, 148 new bacterial phylotypes, 1.2 million previously unknown genes, and 782 new rhodopsin-like photo-receptors.

Thus research in marine microbiology, if properly approached, may reveal a wealth of information in the future. In the research program, the researchers are interested in converting non-cellulose users into cellulose utilizers. They are doing this by transferring Clostridium cellulovorans cellulosome genes into a non-cellulose degrader such as *Bacillus subtilis.* They recently succeeded in constructing a *B. subtilis* strain that is capable of making minicellulosomes. By use of this strategy, it appears possible to convert many microorganisms into cellulose utilizers. This could reduce the cost of production of such items as ethanol, amino acids, and organic acids since relatively inexpensive cellulosic biomass can be used as the starting carbon substrate. Furthermore, they developed an "Intercellular Complementation" system in which one B. subtilis strain expresses and secretes a mini-scaffoldin and another expresses and secretes an endoglucanase EngE.

By growing a mixed culture of the two *B. subtilis* strain, the mini-scaffoldin has interacted with EngE in the culture medium to form minicellulosomes. This strategy can be used to produce designer cellulosomes. Thus many novel approaches will be used in the future that take advantage of microbial biotechnology. This should lead to the development of many useful technologies for health care, for the food and pharmaceutical industries, and for the alleviation of many of the environmental problems facing the world.

References

Fellows, P., 1997, *Traditional Foods*, Intermediate Technology Publications, UK.

Dirar, H., 1993, *The Indigenous Fermented Foods of the Sudan: A study in African Food and Nutrition*, CAB International, UK.

Ranken, M.D., Kill, R.C. and Baker, C.G.J., 1997, *Food Industries Manual,* Blackie Academic and Professional, UK.

Rose, A., 1961, *Industrial Microbiology*, Butterworths, UK.

3

Fermentation Technology

Agricultural crops are processed for many different reasons. These range from the removal of anti-nutritional components and increasing the storage life of the final product to adding value to increase both employment and income generating opportunities. Fermentation is one of the most ancient and most important food processing technologies. However scientists and policy makers have neglected this area, particularly traditional fermented products from developing countries.

Development of Fermentation Technologies

Fermentation is one of the oldest forms of food preservation technologies in the world. Indigenous fermented foods such as bread, cheese and wine, have been prepared and consumed for thousands of years and are strongly linked to culture and tradition, especially in rural households and village communities.

The development of fermentation technologies is lost in the mists of history. Anthropologists have suggested that it was the production of alcohol that motivated primitive people to settle down and become agriculturists. Some even think the consumption of fermented food is pre-human. The first fermented foods consumed probably were fermented fruits. Hunter-gatherers would have consumed fresh fruits but at times of scarcity would have eaten rotten and fermented fruits. Repeated consumption would have led to the development of the taste for fermented fruits. There is reliable information that fermented drinks were being produced over 7,000 years ago in Babylon (now Iraq), 5,000 years ago in Egypt, 4,000 years ago in Mexico and 3,500 years ago in Sudan.

Bread-making probably originated in Egypt over 3,500 years ago. Several triangular loaves of bread have been found in ancient tombs. Fermentation of milk started in many places with evidence of fermented products in use in Babylon over 5,000 years ago. There is also evidence of fermented meat products being produced for King Nebuchadnezer of Babylon. China is thought to be the birth-place of fermented vegetables and the use of *Aspergillus* and *Rhizopus* moulds to make food. The book called "Shu-Ching" written in the Chou dynasty in China (1121-256 BC) refers to the use of "chu" a fermented grain product.

Knowledge about traditional fermentation technologies has been handed down from parent to child, for centuries. These fermented products have been adapted over generations; some products and practices no doubt fell by the wayside. Those that remain today have not only survived the test of time but also more importantly are appropriate to the technical, social and economic conditions of the region.

Fermentation is a relatively efficient, low energy preservation process which increases the shelf life and decreases the need for refrigeration or other form of food preservation technology. It is therefore a highly appropriate technique for use in developing countries and remote areas where access to sophisticated equipment is limited. Fermented foods are popular throughout the world and in some regions make a significant contribution to the diet of millions of individuals.

In Asia the preparation of fermented foods is a widespread tradition. The fermented products supply protein, minerals and other nutrients that add variety and nutritional fortification to otherwise starchy, bland diets. For instance Soy sauce is consumed throughout the world and is a fundamental ingredient in diets from Indonesia to Japan. Over one billion litres are produced each year in Japan alone. "*Gundruk*" which is a fermented and dried vegetable product is very important for ensuring food security for many Nepali communities especially in remote areas. It is served as a side dish with the main meal and is also used as an appetiser in the bland, starchy diet. The annual production of *gundruk* in Nepal is estimated at 2,000 tons. *Gundruk* is an important source of minerals particularly during the off-season when the diet consists primarily of starchy tubers and maize, which tend to be low in minerals. In Africa fermented cassava products (like *gari* and *fufu*) are a major component of the diet of more than 800 million people and in some areas these products constitute over 50% of the diet.

Benefits of Fermenting Fruits And Vegetables

Fermenting fruits and vegetables can bring many benefits to people in developing countries. Fermented foods play an important role in providing food security, enhancing livelihoods and improving the nutrition and social well being of millions of people around the world, particularly the marginalised and vulnerable.

Improving Food Security

Eight hundred million people do not have enough food to eat. If we include those not free from hunger the figure rises to 1.2 billion people. This is one fifth of the World's population. A further two billion people are deficient in one or more micro-nutrients. In the seventies, food security was viewed mainly in terms of food supply at the global and national levels. Since then there has been a major shift in understanding of food security with more emphasis on access to food rather than purely on production. The Food and Agriculture Organisation of the United Nations (FAO), amongst other influential organisations, has recognised that the problem of food security cannot be tackled in isolation. Moreover that it is an integral component of other development issues. FAO highlights the fact that the world food insecurity problem is a result of undemocratic and inequitable distribution of and access to resources rather than a problem of global food production.

Fermentation technologies play an important role in ensuring the food security of millions of people around the world, particularly marginalised and vulnerable groups. This is achieved through improved food preservation, increasing the range of raw materials that can be used to produce edible food products and removing anti- nutritional factors to make food safe to eat.

Food Preservation

Fermentation is a cheap and energy efficient means of preserving perishable raw materials. When harvested, fruit and vegetables, undergo rapid deterioration, especially in the humid tropics where the prevailing environmental conditions accelerate the process of decomposition. There are several options for preserving fresh fruit and vegetables including drying, freezing, canning and pickling. However many of these are inappropriate for use on the small-scale in developing countries. For instance the canning of vegetables at the small-scale has serious food safety implications and contamination with botulism is a possibility. Freezing of fruits and vegetables is not economically viable at the small-scale. Fermentation requires very little sophisticated equipment, either to carry out the fermentation or for subsequent storage of the fermented product. It is a technique that has been employed for generations to preserve food for consumption at a later date and to improve food security. There are examples from around the world of the role fermented foods have played in preserving food to enhance food security.

Salvaging Waste Foods

Fermentation can salvage waste food which otherwise would not be usable as food by changing the consistency of the product and making it digestible. This increases the range of raw materials available as food.

Removal of Anti-nutritional Factors

Many fruits and vegetables contain naturally occurring toxins and anti-nutritional compounds. These can be removed or detoxified by the action of micro-organisms during fermentation. For instance the fermentation process that produces the Sudanese product *Kawal*removes the toxins from the leaves of *Cassia obtusifolia* and fermentation is an important step in ensuring that cassava is safe to eat.

Increasing Income and Employment

The production of fermented fruit and vegetable products provides income and employment to millions of people around the world.

Food processing is probably the most important source of income and employment in Africa, Asia and Latin America. The Food and Agriculture Organisation of the United Nations has stated that value added through marketing and processing raw products can be much greater than the value of primary production. For instance in sub-Saharan Africa more than 60% of the workforce is employed in the small scale food processing sector, and between one third and two thirds of value added manufacturing is based on agricultural raw materials. This is particularly important as agriculture and the formal sector are unable to absorb the growing labour force in many countries.

Fermented foods are popular throughout the world and the production of fermented food products is important in many countries in providing income and employment.

In Africa, fermented cassava products (like *Gari* and *Fufu)* are a major component of the diet of more than 800 million people and in some parts of Africa it constitutes over 50% of the diet. In Asia the preparation of fermented foods is a widespread tradition. *Kimchi* (a fermented cabbage product) is the major food product of Korea. Soy sauce (a fermented legume product) is economically important from Indonesia to Japan. Over a billion litres are produced each year in Japan alone. Over 2000 million litres are produced each year in Korea and over 150 million litres in Taiwan. *Miso* (a fermented legume product) is also very important in Asia with over 560,000 tons produced a year in Japan alone. In Latin America, fermented cereal products, alcoholic drinks and fermented milk products are three of the most important sectors of the economy.

Improving Nutrition

The optimum health and nutrition of individuals is dependent upon a regular supply of food and a balanced diet. When diets are sub-optimal, the individual's capacity for work and achievements are greatly reduced. The most vulnerable groups are women, children and weaning infants. Availability of food, dietary restrictions and taboos, misconceptions, limited time available for feeding or eating compound to create a group of individuals who are nutritionally disadvantaged. Approximately 30% of women consume less than their daily requirements of energy and at least 40% of women world-wide suffer from iron-deficiency anaemia. Fermentation can enhance the nutritional value of a food product though increased vitamin levels and improved digestibility.

Vitamins

Fermentation processes can result in increased levels of vitamins in the final product. *Saccharomyces cerevisiae* is able to concentrate large quantities of thiamin, nicotinic acid and biotin and thus form enriched products.

— Sorghum beer in Southern Africa contains relatively high levels of riboflavin and nicotinic acid, which are important for people consuming a high maize diet. Pellagra (a vitamin deficiency disease associated with high maize diets) is unusual in communities in which sorghum beer is consumed. Even children benefit from consuming the dregs which contain relatively little alcohol but are rich in vitamins.

— Palm wine in West Africa is high in vitamin B12, which is very important for people with low meat intake, and who subsist primarily on a vegetarian diet.

— *Pulque* (a fermented plant sap) is an important source of vitamins for the economically deprived in Mexico. The fermentation process involved in *Pulque* production increases its vitamin content. For instance the vitamin content (milligrams of vitamins per 100g of product) of pulque increases from 5 to 29 for thiamine, 54 to 515 for niacin and 18 to 33 for riboflavin during fermentation.

— *Idli* (a lactic acid bacteria fermented product consumed in India) is high in thiamine and riboflavin.

Digestibility

Micro-organisms contain certain enzymes, such as cellulases, which are incapable of being synthesised by humans. Microbial cellulases hydrolyse cellulose into sugars which are then readily digestible by humans. Similarly pectinases soften the texture of foods and liberates sugars for digestion. Fermented foods are often more easily digestible than unfermented foods.

Lactic acid fermented weaning foods are traditionally produced in developing countries, both to improve the safety of the food and to improve its digestibility. Starchy porridges are commonly fed to weaning infants in developing countries. The consistency of these gruels, combined with the small capacity of the infants stomach, means that it is physically impossible for the child to consume adequate energy to meet its high demands. By acidifying the porridge through lactic acid fermentation, starch is hydrolysed into shorter chains of glucose and dextrose, which reduce the viscosity of the porridge and increase its energy density. Thus the child is more able to meet its energy requirements.

Medicinal benefits

There are many traditional beliefs about the medicinal properties of fermented food products. The Fur ethnic group in Sudan strongly believe that the consumption of fermented foods protects them from disease. *Koumiss* (a fermented milk product in Russia) has been used to treat tuberculosis. *Pulque* (a fermented fruit sap) is felt to have medicinal properties in Mexico.

Fermentation is a traditional method of reducing the microbial contamination of porridges in Kenya A study in Tanzania has shown that children fed with fermented gruels had a 33% lower incidence of diarrhoea than those fed unfermented gruels, owing to the inhibition of pathogenic bacteria by lactic acid forming bacteria.

Improving Cultural and Social Well-being

Fermentation can improve the flavour and appearance of food. One important area is the creation of meat-like flavour. Over the years, Sudanese women have developed products to replace meat in their diets. These include "*kawal*", fermented wild legume leaves, "*sigda*"(fermented sesame press-cake) and "*furundu*" (fermented red sorrel seeds). The strong flavours of fermented food products can enhance a dull diet. Fermented vegetables such as pickles, *gundruk* and sauerkraut are used as condiments to enhance the overall flavour of the meal. A small amount of pickle can make a bland starchy diet (like dahl and rice in Asia) much more appealing.

Basic Principles of Fermentation

Numerous fermented foods are consumed around the world. Each nation has its own types of fermented food, representing the staple diet and the raw ingredients available in that particular place. Although the products are well know to the individual, they may not be associated with fermentation. Indeed, it is likely that the methods of producing many of the worlds fermented foods are unknown and came about by chance. Some of the more obvious fermented fruit and vegetable products are the alcoholic beverages - beers and wines. However, several more fermented fruit and vegetable products arise from lactic acid fermentation and are extremely important in meeting the nutritional requirements of a large proportion of the worlds population. Table 1 contains examples of fermented fruit and vegetable products from around the world.

Organisms Responsible for Food Fermentations

The most common groups of micro-organisms involved in food fermentations are:

- Bacteria
- Yeasts
- Moulds

Table 1. Fermented foods from around the world

Name and region	*Type of product*
Indian sub-continent	
Achar	Pickled fruit and vegetables
Gundruk	Fermented dried vegetable
Pickle	
South East Asia	
Asinan, Mangga, Dalok, Kiam-chai, Kiam-cheyi, Kong-chai, Naw-mai-dong, Pak-siam-dong, Paw-tsay, Phak-dong, Phonlami-dong, Sajur asin, Sambal tempo-jak, Santol, Si-sek-chai, Sunki, Tang-chai, Tempoyak, Vanilla	Pickled fruit and vegetables
Bai-ming, Leppet-so, Miang	Fermented tea leaves
Nata de coco, Nata de pina	Fermented fruit juice
East Asia	
Bossam-kimchi, Chonggak-kimchi, Dan moogi, Dongchimi, Kachdoo kigactuki, Kakduggi, Kimchi, Mootsanji, Muchung-kimchi, Oigee, Oiso baegi, Tongbaechu-kimchi, Tongkimchi, Totkal kimchi,	Fermented in brine
Cha-ts'ai, Hiroshimana, Jangagee, Nara senkei, Omizuke, Pow tsai, Red in snow, Seokbakji, Shiozuke, Szechwan cabbage, Tai-tan tsoi, Takana, Tsa Tzai, Tsu, Umeboshi, Wasabi-zuke, Yen tsai	Pickled fruit and vegetables
Hot pepper sauce	
Africa	
Fruit vinegar	Vinegar
Hot pepper sauce	
Lamoun makbouss, Msir, Mslalla, Olive	Pickled fruit and vegetables
Oilseeds, Ogili, Ogiri, Hibiscus seed	Fermented fruit and vegetable seeds
Wines	Fermented fruits
Americas	
Cucumber pickles, Olives, Sauerkraut,	Pickled fruit and vegetables
Lupin seed, Oilseeds	Pickled oilseed
Vanilla, Wines	Fermented fruit and vegetable
Middle East	
Kushuk	Fermented fruit and vegetables
Lamoun Mekhalel, Olives, Torshi, Tursu	Pickled fruit and vegetables
Wines	Fermented fruits
Europe and World	
Mushrooms, Yeast	Moulds
Olives, Sauerkohl, Sauerruben	Pickled fruit and vegetables
Grape vinegar, Wine vinegar	Vinegar
Wines, Citron	Fermented fruits

Bacteria

Several bacterial families are present in foods, the majority of which are concerned with food spoilage. As a result, the important role of bacteria in the fermentation of foods is often overlooked. The most important bacteria in desirable food fermentations are the*lactobacillaceae* which have the ability to produce lactic acid from carbohydrates. Other important bacteria, especially in the fermentation of fruits and vegetables, are the acetic acid producing *acetobacter* species.

Yeasts

Yeasts and yeast-like fungi are widely distributed in nature. They are present in orchards and vineyards, in the air, the soil and in the intestinal tract of animals. Like bacteria and moulds, yeasts can have beneficial and non-beneficial effects in foods. The most beneficial yeasts in terms of desirable food fermentation are from the *Saccharomyces* family, especially *S. cerevisiae*. Yeasts are unicellular organisms that reproduce asexually by budding. In general, yeasts are larger than most bacteria. Yeasts play an important role in the food industry as they produce enzymes that favour desirable chemical reactions such as the leavening of bread and the production of alcohol and invert sugar.

Moulds

Moulds are also important organisms in the food industry, both as spoilers and preservers of foods. Certain moulds produce undesirable toxins and contribute to the spoilage of foods. The *Aspergillus* species are often responsible for undesirable changes in foods. These moulds are frequently found in foods and can tolerate high concentrations of salt and sugar. However, others impart characteristic flavours to foods and others produce enzymes, such as amylase for bread making. Moulds from the genus *Penicillium* are associated with the ripening and flavour of cheeses. Moulds are aerobic and therefore require oxygen for growth. They also have the greatest array of enzymes, and can colonise and grow on most types of food. Moulds do not play a significant role in the desirable fermentation of fruit and vegetable products.

When micro-organisms metabolise and grow they release by-products. In food fermentations the by-products play a beneficial role in preserving and changing the texture and flavour of the food substrate. For example, acetic acid is the by-product of the fermentations of some fruits. This acid not only affects the flavour of the final product, but more importantly has a preservative effect on the food. For food fermentations, micro-organisms are often classified according to these by-products. The fermentation of milk to yoghurt involves a specific group of bacteria called the lactic acid bacteria (*Lactobacillus* species). This is a general name attributed to those bacteria which produce lactic acid as they grow. Acidic foods are less susceptible to spoilage than neutral or alkaline foods and hence the acid helps to preserve the product. Fermentations also result in a change in texture. In the case of milk, the acid causes the precipitation of milk protein to a solid curd.

Enzymes

The changes that occur during fermentation of foods are the result of enzymic activity. Enzymes are complex proteins produced by living cells to carry out specific biochemical reactions. They are known as catalysts since their role is to initiate and control reactions, rather than being used in a reaction. Because they are proteinaceous in nature, they are sensitive to fluctuations in temperature, pH, moisture content, ionic strength and concentrations of substrate and inhibitors. Each enzyme has requirements at which it will operate most efficiently. Extremes of temperature and pH will denature the protein and destroy enzyme activity. Because they are so sensitive, enzymic reactions can easily be controlled by slight adjustments to temperature, pH or other reaction conditions. In the food industry, enzymes have several roles - the liquefaction and saccharification of starch, the conversion of sugars and the modification of proteins. Microbial enzymes play a role in the fermentation of fruits and vegetables.

Nearly all food fermentations are the result of more than one micro-organism, either working together or in a sequence. For example, vinegar production is a joint effort between yeast and acetic acid forming bacteria. The yeast convert sugars to alcohol, which is the substrate required by the *acetobacter* to produce acetic acid. Bacteria from different species and the various micro-organisms - yeast and moulds -all have their own preferences for growing conditions, which are set within narrow limits. There are very few pure culture fermentations. An organism that initiates fermentation will grow there until it's by-products inhibit further growth and activity. During this initial growth period, other organisms develop which are ready to take over when the conditions become intolerable for the former ones.

In general, growth will be initiated by bacteria, followed by yeasts and then moulds. There are definite reasons for this type of sequence. The smaller micro-organisms are the ones that multiply and take up nutrients from the surrounding area most rapidly. Bacteria are the smallest of micro-organisms, followed by yeasts and moulds. The smaller bacteria, such as *Leuconostoc* and *Streptococcus* grow and ferment more rapidly than their close relations and are therefore often the first species to colonise a substrate.

Desirable Fermentation

It is essential with any fermentation to ensure that only the desired bacteria, yeasts or moulds start to multiply and grow on the substrate. This has the effect of suppressing other micro-organisms which may be either pathogenic and cause food poisoning or will generally spoil the fermentation process, resulting in an end-product which is neither expected or desired. An everyday example used to illustrate this point is the differences in spoilage between pasteurised and unpasteurised milk. Unpasteurised milk will spoil naturally to produce a sour tasting product which can be used in baking to improve the texture of certain breads. Pasteurised milk, however, spoils (non-desirable fermentation) to produce an unpleasant product which has to be disposed of. The reason for this difference is that pasteurisation (despite being a very important process to destroy pathogenic micro-organisms) changes the micro-organism environment and if pasteurised milk is kept unrefrigerated for some time, undesirable micro-organisms start to grow and multiply before the desirable ones. In the case of unpasteurised milk, the non-pathogenic lactic acid bacteria start to grow and multiply at a greater rate that any pathogenic bacteria. Not only do the larger numbers of lactic acid bacteria compete more successfully for the available nutrients, but as they grow they produce lactic acid which increases the acidity of the substrate and further suppresses the bacteria which cannot tolerate an acid environment.

Most food spoilage organisms cannot survive in either alcoholic or acidic environments. Therefore, the production of both these end products can prevent a food from spoilage and extend the shelf life. Alcoholic and acidic fermentations generally offer cost effective methods of preserving food for people in developing countries, where more sophisticated means of preservation are unaffordable and therefore not an option.

The principles of microbial action are identical both in the use of micro-organisms in food preservation, such as through desirable fermentations, and also as agents of destruction via food spoilage. The type of organisms present and the environmental conditions will determine the nature of the reaction and the ultimate products. By manipulating the external reaction conditions, microbial reactions can be controlled to produce desirable results. There are several means of altering the reaction environment to encourage the growth of desirable organisms. These are discussed below.

Manipulation of Microbial Growth and Activity

There are six major factors that influence the growth and activity of micro-organisms in foods. These are moisture, oxygen concentration, temperature, nutrients, pH and inhibitors. The food supply available to the micro-organisms depends on the composition of the food on which they grow. All micro-organisms differ in their ability to metabolise proteins, carbohydrates and fats. Obviously, by manipulating any of these six factors, the activity of micro-organisms within foods can be controlled.

Moisture

Water is essential for the growth and metabolism of all cells. If it is reduced or removed, cellular activity is decreased. For example, the removal of water from cells by drying or the change in state of water (from liquid to solid) affected by freezing, reduces the availability of water to cells (including microbial cells) for metabolic activity. The form in which water exists within the food is important as far as microbial activity is concerned. There are two types of water - free and bound. Bound water is present within the tissue and is vital to all the physiological processes within the cell. Free water exists in and around the tissues and can be removed from cells without seriously interfering with the vital processes. Free water is essential for the survival and activity of micro-organisms. Therefore, by removing free water, the level of microbial activity can be controlled. The amount of water available for micro-organisms is referred to as the water activity (a_w). Pure water has a water activity of 1.0. Bacteria require more water than yeasts, which require more water than moulds to carry out their metabolic activities. Almost all microbial activity is inhibited below a_w of 0.6. Most fungi are inhibited below a_w of 0.7, most yeasts are inhibited below a_w of 0.8 and most bacteria below a_w 0.9. Naturally, there are exceptions to these guidelines and several species of micro-organism can exist outside the stated range. The water activity of foods can be changed by altering the amount of free water available. There are several ways to achieve this – drying to remove water; freezing to change the state of water from liquid to solid; increasing or decreasing the concentration of solutes by adding salt or sugar or other hydrophylic compounds (salt and sugar are the two common additives used for food preservation). Addition of salt or sugar to a food will bind free water and so decrease the a_w. Alternatively, decreasing the concentration will increase the amount of free water and in turn the a_w. Manipulation of the a_w in this manner can be used to encourage the growth of favourable micro-organisms and discourage the growth of spoilage ones.

Oxidation-Reduction potential

Oxygen is essential to carry out metabolic activities that support all forms of life. Free atmospheric oxygen is utilised by some groups of micro-organisms, while others are able to metabolise the oxygen which is bound to other compounds such as carbohydrates. This bound oxygen is in a reduced form.

Micro-organisms can be broadly classified into two groups - aerobic and anaerobic. Aerobes grow in the presence of atmospheric oxygen while anaerobes grow in the absence of atmospheric oxygen. In the middle of these two extremes are the facultative anaerobes which can adapt to the prevailing conditions and grow in either the absence or presence of atmospheric oxygen. Microaerophilic organisms grow in the presence of reduced amounts of atmospheric oxygen. Thus, controlling the availability of free oxygen is one means of controlling microbial activity within a food. In aerobic fermentations, the amount of oxygen present is one of the limiting factors. It determines the type and amount of biological product obtained, the amount of substrate consumed and the energy released from the reaction.

Temperature

Temperature affects the growth and activity of all living cells. At high temperatures, organisms are destroyed, while at low temperatures, their rate of activity is decreased or suspended. Micro-organisms can be classified into three distinct categories according to their temperature preference.

Nutritional requirements

The majority of organisms are dependent on nutrients for both energy and growth. Organisms vary in their specificity towards different substrates and usually only colonise foods which contain the substrates they require. Sources of energy vary from simple sugars to complex carbohydrates and proteins. The energy requirements of micro-organisms are very high. Limiting the amount of substrate available can check their growth.

Hydrogen ion concentration (pH)

The pH of a substrate is a measure of the hydrogen ion concentration. A food with a pH of 4.6 or less is termed a high acid or acid food and will not permit the growth of bacterial spores. Foods with a pH above 4.6. are termed low acid and will not inhibit the growth of bacterial spores. By acidifying foods and achieving a final pH of less than 4.6, most foods are resistant to bacterial spoilage.

The optimum pH for most micro-organisms is near the neutral point (pH 7.0). Certain bacteria are acid tolerant and will survive at reduced pH levels. Notable acid-tolerant bacteria include *the Lactobacillus* and *Streptococcus* species, which play a role in the fermentation of dairy and vegetable products. Moulds and yeasts are usually acid tolerant and are therefore associated with spoilage of acidic foods.

Micro-organisms vary in their optimal pH requirements for growth. Most bacteria favour conditions with a near neutral pH (7). Yeasts can grow in a pH range of 4 to 4.5 and moulds can grow from pH 2 to 8.5, but favour an acid pH. The varied pH requirements of different groups of micro-organisms is used to good effect in fermented foods where successions of micro-organisms take over from each other as the pH of the environment changes. For instance, some groups of micro-organisms ferment sugars so that the pH becomes too low for the survival of those microbes. The acidophilic micro-organisms then take over and continue the reaction. The affinity for different pH can also be used to good effect to occlude spoilage organisms.

Inhibitors

Many chemical compounds can inhibit the growth and activity of micro-organisms. They do so by preventing metabolism, denaturation of the protein or by causing physical damage to the cell. The production of substrates as part of the metabolic reaction also acts to inhibit microbial action.

Controlled Fermentation

Controlled fermentations are used to produce a range of fermented foods, including sauerkraut, pickles, olives, vinegar, dairy and other products. Controlled fermentation is a form of food preservation since it generally results in a reduction of acidity of the food, thus preventing the growth of spoilage micro-organisms. The two most common acids produced are lactic acid and acetic acid, although small amounts of other acids such as propionic, fumaric and malic acid are also formed during fermentation.

It is highly probable that the first controlled food fermentations came into existence through trial and error and a need to preserve foods for consumption later in the season. It is possible that the initial attempts at preservation involved the addition of salt or seawater. During the removal of the salt prior to consumption, the foods would pass through stages favourable to acid fermentation. Although the process worked, it is likely that the causative agents were unknown. Today, there are numerous examples of controlled fermentation for the preservation and processing of foods. However, only a few of these have been studied in any detail - these include sauerkraut, pickles, kimchi, beer, wine and vinegar production. Although the general principles and processes for many of the fermented fruit and vegetable products are the same -relying mainly on lactic acid and acetic acid forming bacteria, yeasts and moulds, the reactions have not been quantified for each product. The reactions are usually very complex and involve a series of micro-organisms, either working together or in succession to achieve the final product.

Yeast Fermentations

A yeast is a unicellular fungus which reproduces asexually by budding or division, especially the genus *Saccharomyces* which is important in food fermentations. Yeasts and yeast-like fungi are widely distributed in nature. They are present in orchards and vineyards, in the air, the soil and the intestinal tract of animals. Like bacteria and moulds, they can have beneficial and non-beneficial effects in foods. Most yeasts are larger than most bacteria. The most well known examples of yeast fermentation are in the production of alcoholic drinks and the leavening of bread. For their participation in these two processes, yeasts are of major importance in the food industry.

Some yeasts are chromogenic and produce a variety of pigments, including green, yellow and black. Others are capable of synthesising essential B group vitamins.

Although there is a large diversity of yeasts and yeast-like fungi, (about 500 species), only a few are commonly associated with the production of fermented foods. They are all either ascomycetous yeasts or members of the genus *Candida*. Varieties of the *Saccharomyces cervisiae* genus are the most common yeasts in fermented foods and beverages based on fruit and vegetables. All strains of this genus ferment glucose and many ferment other plant derived carbohydrates such as sucrose, maltose and raffinose. In the tropics,*Saccharomyces pombe* is the dominant yeast in the production of traditional fermented beverages, especially those derived from maize and millet.

Conditions Necessary for Fermentation

Most yeasts require an abundance of oxygen for growth, therefore by controlling the supply of oxygen, their growth can be checked. In addition to oxygen, they require a basic substrate such as sugar. Some yeasts can ferment sugars to alcohol and carbon dioxide in the absence of air but require oxygen for growth. They produce ethyl alcohol and carbon dioxide from simple sugars such as glucose and fructose.

In conditions of excess oxygen (and in the presence of acetobacter) the alcohol can be oxidised to form acetic acid. This is undesirable if the end product is a fruit alcohol, but is a technique employed for the production of fruit vinegars.

Yeasts are active in a very broad temperature range - from 0 to 50° C, with an optimum temperature range of 20° to 30° C.

The optimum pH for most micro-organisms is near the neutral point (pH 7.0). Moulds and yeasts are usually acid tolerant and are therefore associated with the spoilage of acidic foods. Yeasts can grow in a pH range of 4 to 4.5 and moulds can grow from pH 2 to 8.5, but favour an acid pH.

In terms of water requirements, yeasts are intermediate between bacteria and moulds. Bacteria have the highest demands for water, while moulds have the least need. Normal yeasts require a minimum water activity of 0.85 or a relative humidity of 88%.

Yeasts are fairly tolerant of high concentrations of sugar and grow well in solutions containing 40% sugar. At concentrations higher than this, only a certain group of yeasts – the osmophilic type – can survive. There are only a few yeasts that can tolerate sugar concentrations of 65-70% and these grow very slowly in these conditions. Some yeasts – for example the *Debaromyces* - can tolerate high salt concentrations. Another group which can tolerate high salt concentrations and low water activity is *Zygosaccharomyces rouxii*, which is associated with fermentations in which salting is an integral part of the process.

Production of Fruit Alcohol

Alcohol and acids are two primary products of fermentation, both used to good effect in the preservation of foods. Several alcohol-fermented foods are preceded by an acid fermentation and in the presence of oxygen and *acetobacter*, alcohol can be fermented to produce acetic acid. Most food spoilage organisms cannot survive in either alcoholic or acidic environments. Therefore, the production of both these end products can prevent a food from undergoing spoilage and extend its shelf life.

Primitive wines and beers have been produced, with the aid of yeasts, for thousands of years, although it was not until about four hundred years ago that micro-organisms associated with the fermentation were observed and identified. It was not until the 1850's that Louis Pasteur demonstrated unequivocally the involvement of yeasts in the production of wines and beers. Since then, the knowledge of yeasts and the conditions necessary for fermentation of wine and beer has increased to the point where pure culture fermentations are now used to ensure consistent product quality. Originally, alcoholic fermentations would have been spontaneous events that resulted from the activity of micro-organisms naturally present. These non-scientific methods are still used today for the home preparation of many of the worlds traditional beers and wines.

Alcoholic drinks fall into two broad categories: wines and beers. Wines are made from the juice of fruits and beers from cereal grains. The principal carbohydrates in fruit juices are soluble sugars; the principal carbohydrate in grains is starch, an insoluble polysaccharide. The yeasts that bring about alcoholic fermentation can attack soluble sugars but do not produce starch-splitting enzymes. Wines can therefore be made by the direct fermentation of the raw material, while the production of beer requires the hydrolysis of starch to yield sugars fermentable by yeast, as a preliminary step.

Raw fruit juice is usually a strongly acidic solution, containing from 10 to 25 percent soluble sugars. Its acidity and high sugar concentration make it an unfavourable medium for the growth of bacteria but highly suitable for yeasts and moulds. Raw fruit juice naturally contains many yeasts, moulds, and bacteria, derived from the surface of the fruit. Normally the yeast used in alcoholic fermentation is a strain of the species *Saccharomyces cerevisiae.*

The fermentation may be allowed to proceed spontaneously, or can be "started" by inoculation with a must that has been previously successfully fermented by *S. cerevisiae* var. *ellipsoideus.* Many modern wineries eliminate the original microbial population of the must by pasteurisation or by treatment with sulphur dioxide. The must is then inoculated with a *starter culture* derived from a pure culture of a suitable strain of wine yeast. This procedure eliminates many of the uncertainties and difficulties of older methods. At the start of the fermentation, the must is aerated slightly to build up a large and vigorous yeast population; once fermentation sets in, the rapid production of carbon dioxide maintains anaerobic conditions, which prevent the growth of undesirable aerobic organisms, such as bacteria and moulds. The temperature of fermentation is usually from 25 to 30°C, and the duration of the fermentation process may extend from a few days to two weeks. As soon as the desired degree of sugar disappearance and alcohol production has been attained, the microbiological phase of wine making is over. Thereafter, the quality and stability of the wine depend very largely on preventing further microbial activity, both during the "aging" in wooden casks and after bottling.

At all stages during its manufacture, fruit juice alcohol is subject to spoilage by undesirable microorganisms. Pasteur, whose descriptions of the organisms responsible and recommendations for overcoming them are still valid today, first scientifically explored the problem of the "diseases" of wines. The most serious aerobic spoilage processes are brought about by film-forming yeasts and acetic acid bacteria, both of which grow at the expense of the alcohol, converting it to acetic acid or to carbon dioxide and water. The chief danger from these organisms arises when access of air is not carefully regulated during aging. Much more serious are the diseases caused by fermentative bacteria, particularly rod-shaped lactic acid bacteria, which utilise any residual sugar and impart a mousy taste to the wine. Such wines are known as turned wines. Since oxygen is unnecessary for the growth of lactic acid bacteria, wine spoilage of this kind can occur even after bottling. These risks of spoilage can be minimised by pasteurisation after bottling.

Grape wine

Grape wine is perhaps the most common fruit juice alcohol. Because of the commercialisation of the product for industry, the process has received most research attention and is documented in detail.

The production of grape wine involves the following basic steps: crushing the grapes to extract the juice; alcoholic fermentation; maltolactic fermentation if desired; bulk storage and maturation of the wine in a cellar; clarification and packaging. Although the process is fairly simple, quality control demands that the fermentation is carried out under controlled conditions to ensure a high quality product.

The distinctive flavour of grape wine originates from the grapes as raw material and subsequent processing operations. The grapes contribute trace elements of many volatile substances (mainly terpenes) which give the final product the distinctive fruity character. In addition, they contribute non-volatile compounds (tartaric and malic acids) which impact on flavour and tannins which give bitterness and astringency. The latter are more prominent in red wines as the tannin components are located in the grape skins.

Although yeasts are the principal organisms involved, filamentous fungi, lactic acid bacteria, acetic acid bacteria and other bacterial groups all play a role in the production of alcoholic fruit products.

Normal grapes harbour a diverse micro-flora, of which the principal yeasts (*Saccharomyces cerevisiae*) involved in desirable fermentation are in the minority. Lactic acid bacteria and acetic acid bacteria are also present. The proportions of each and total numbers present are dependent upon a number of external environmental factors including the temperature, humidity, stage of maturity, damage at harvest and application of fungicides. It is essential to ensure proliferation of the desired species at the expense of the non-desired ones. This is achieved through ensuring fermentation conditions are such to encourage *Saccharomyces* species.

The fermentation may be initiated using a starter culture of *Saccharomyces cerevisiae* – in which case the juice is inoculated with populations of yeast of 10^6 to 10^7 cfu/ml juice. This approach produces a wine of generally expected taste and quality. If the fermentation is allowed to proceed naturally, utilising the yeasts present on the surface of the fruits, the end result is less controllable, but produces wines with a range of flavour characteristics. It is likely that natural fermentations are practiced widely around the world, especially for home production of wine.

During alcoholic fermentation, yeasts are the prominent species. The composition of fruit juice – its acid and sugar level and low pH favour the growth of yeasts and production of ethanol that restricts the growth of bacteria and fungi.

In natural fermentations, there is a progressive pattern of yeast growth. Several species of yeast, including *Kloeckera, Hanseniaspora, Candida* and *Metschnikowia*, are active for the first two to three days of fermentation. The build up of end products (ethanol) is toxic to these yeasts and they die off, leaving *Saccharomyces cerevisiae* to continue the fermentation to the end. *S. cerevisiae* can tolerate much higher levels of ethanol (up to 15% v/v or more) than the other species who only tolerate up to 5 or 8% alcohol. Because of its tolerance of alcohol, *S. cerevisiae* dominates wine fermentation and is the species that has been commercialised for starter cultures.

Traditionally, fermentation was carried out in large wooden barrels or concrete tanks. Modern wineries now use stainless steel tanks as these are more hygienic and provide better temperature control. White wines are fermented at 10 to 18° C for about seven to fourteen days. The low temperature and slow fermentation favours the retention of volatile compounds. Red wines are fermented at 20 to 30°C for about seven days. This higher temperature is necessary to extract the pigment from the grape skins.

Factors affecting wine fermentation

There are several variables which can affect the fermentation process and final quality of wine. Factors which are most important to control are:

— the clarification and pre-treatment of juice

— chemical composition of the juice

— temperature of the fermentation

— the influences of other micro-organisms.

Clarification and pre-treatment of juice. Excessive clarification removes many of the natural yeasts and flora. This is beneficial if a tightly controlled induced fermentation is desired, but less so if the fermentation is a natural one. Long periods of settling out however, encourage the growth of natural flora, which can contribute to the fermentation.

Chemical composition of juice. The main consituents of grape juice are glucose (75 to 150 g/l), fructose (75 to 150 g/l), tartaric acid (2 to 10 g/l), malic acid (1 to 8 g/l) and free amino acids (0.2 to 2.5 g/l). The main reaction is the fermentation of glucose and fructose to ethanol and carbon dioxide. However, the presence of nitrogenous and sulphurous products also contributes to the fermentation. The addition of sulphur dioxide to the juice delays the growth of yeast, but does not necessarily inhibit growth of the non-Saccharomyces strains. Fruits generally contain sufficient substrates - soluble sugars - for the yeast to ferment and convert into an acceptable concentration of alcohol. Sugar can be added to fruit juices with a low sugar content, to increase the amount of fermentable substrate.

Temperature. Temperature has an impact on the growth and activity of different strains of yeast. At temperatures of 10 to 15° C, the non-Saccharomyces species have an increased tolerance to alcohol and therefore have the potential to contribute to the fermentation.

Influence of other micro-organisms. Other micro-organisms have the potential to influence wine production at all stages of the process. Prior to harvest, yeasts grow on the surface of grapes. Fungicides are used in an attempt to control their growth, but these disturb the natural balance of flora, thus making it difficult to carry out a 'natural' fermentation. Overuse of fungicides can lead to the development of resistant strains of yeast which have the potential to produce toxins which destroy the desirable yeast species. These yeasts are known as 'killer' strains. Other microbes have further chances to influence the fermentation during the clarification process, after fermentation and during maturation and bottling when*acetobacter* species can oxidise the alcohol and produce acetic acid.

About two to three weeks after the alcoholic fermentation is finished wines often undergo a malo-lactic fermentation. This occurs naturally and lasts for about four weeks. It is a lactic acid fermentation, initiated by lactic acid bacteria resident in the wine. Inoculating the fermented wine with cultures of *Leuconostoc oenos* can start the process if it is desired. The main reaction of these bacteria is the decarboxylation of L-malic acid to L-lactic acid, which decreases the acidity of the wine and increases its pH by about 0.3 to 0.5 units. Wines produced from grapes grown in colder climates tend to have a higher concentration of malic acid and a lower pH (3.0 to 3.5) and the taste benefits from this slight decrease in acidity. The benefits of this process are that it imparts a more mellow flavour to the wine. The growth of malo-lactic bacteria also contributes to the taste of the wine. Wines that have undergone a malo-lactic fermentation appear to be less susceptible to any further damage from other bacteria. This could be because *L. oenos* has used up all available substrate, or it may have secreted bacteriocins which prevent the growth of other species. Although the malo-lactic fermentation seems to be a useful process, not all wines benefit from it. Wines produced from grapes in warmer climates tend to be less acidic (pH > 3.5) and a further reduction in acidity may have adverse effects on the quality of the wine. Decreasing the acidity also increases the pH to values which can allow spoilage organisms to multiply. It is difficult to prevent the malo-lactic fermentation from taking place naturally, especially later on after the wine has been bottled. In low acid wines, the acidity may be adjusted after this fermentation has taken place. The malo-lactic fermentation can be prevented by controlling several factors: the wine pH (< 3.2); ethanol content (> 14%) and levels of sulphur dioxide (>50 mg/l). The bacteriocin nisin can also be used to control the growth of malo-lactic bacteria. However the subtle blend of aromas and flavours that contribute to the final taste may be lost by such stringent control.

The conversion of malic acid to lactic acid is one of the main reactions carried out by wine lactic acid bacteria. *L oenos* needs to be present in significant numbers (greater than 10^6 cfu/ml) for the reaction to take place at a suitable pace. The bacteria use residual pentose and hexose sugars in the wine as a substrate for growth. The main reaction is the deacidification (or decarboxylation) of malic acid. In addition to this, the by-products of the reaction impart flavours and aromas to the wine.

During storage, wines are prone to non-desirable microbial changes. Yeasts, lactic acid bacteria, acetic acid bacteria and fungi can all spoil or taint wines after the fermentation process is completed. The changes that occur are increased acidification through the formation of acetic and other acids from alcohol; increased carbonation through a secondary fermentation of residual sugars and flavour changes through the metabolism of numerous compounds.

Bacterial Fermentations

Bacteria are "a large group of unicellular or multi-cellular organisms lacking chlorophyll, with a simple nucleus, multiplying rapidly by simple fission, some species developing a highly resistant resting (spore) phase; some species reproduce sexually, and some are motile. In shape they are spherical, rodlike, spiral, or filamentous. They occur in air, water, soil, rotting organic material, animals and plants. Saprophytic forms are more numerous than parasites. A few forms are autotrophic".

There are several bacterial families present in foods, the majority of which are concerned with food spoilage. The important role of bacteria in the fermentation of foods is often overlooked.

Lactic Acid Bacteria

The lactic acid bacteria are a group of Gram positive bacteria, non-respiring, non-spore forming, cocci or rods, which produce lactic acid as the major end product of the fermentation of carbohydrates. They are the most important bacteria in desirable food fermentations, being responsible for the fermentation of sour dough bread, sorghum beer, all fermented milks, cassava (to produce *gari* and *fufu*) and most "pickled" (fermented) vegetables. Historically, bacteria from the genera *Lactobacillus, Leuconostoc, Pediococcus* and*Streptococcus* are the main species involved. Several more have been identified, but play a minor role in lactic fermentations. Lactic acid bacteria were recently reviewed by Axelsson.

Lactic acid bacteria carry out their reactions - the conversion of carbohydrate to lactic acid plus carbon dioxide and other organic acids - without the need for oxygen. They are described as microaerophilic as they do not utilise oxygen. Because of this, the changes that they effect do not cause drastic changes in the composition of the food. Some of the family are homofermentative, that is they only produce lactic acid, while others are heterofermentative and produce lactic acid plus other volatile compounds and small amounts of alcohol.*Lactobacillus acidophilus, L. bulgaricus, L. plantarum, L. caret, L. pentoaceticus, L brevis* and *L. thermophilus* are examples of lactic acid-producing bacteria involved in food fermentations. All species of lactic acid bacteria have their own particular reactions and niches, but overall, *L. plantarum* – a homofermenter -produces high acidity in all vegetable fermentations and plays the major role. All lactic acid producers are non-motile gram positive rods that need complex carbohydrate substrates as a source of energy. The lactic acid they produce is effective in inhibiting the growth of other bacteria that may decompose or spoil the food. Because the whole group are referred to as 'lactic acid bacteria' it might appear that the reactions they carry out are very simple, with the production of one substrate. This is far from the truth. The lactic acid bacteria are a diverse group of organisms with a diverse metabolic capacity. This diversity makes them very adaptable to a range of conditions and is largely responsible for their success in acid food fermentations.

Despite their complexity, the whole basis of lactic acid fermentation centres on the ability of lactic acid bacteria to produce acid, which then inhibits the growth of other non-desirable organisms. All lactic acid producers are micro-aerophilic, that is they require small amounts of oxygen to function. Species of the genera *Streptococcus* and *Leuconostoc* produce the least acid. Next are the heterofermentative species of *Lactobacillus* which produce intermediate amounts of acid, followed by the *Pediococcus* and lastly the homofermenters of the *Lactobacillus* species, which produce the most acid. Homofermenters, convert sugars primarily to lactic acid, while heterofermenters produce about 50% lactic acid plus 25% acetic acid and ethyl alcohol and 25% carbon dioxide. These other compounds are important as they impart particular tastes and aromas to the final product. The heterofermentative lactobacilli produce mannitol and some species also produce dextran.

Leuconostoc mesenteroides is a bacterium associated with the sauerkraut and pickle fermentations. This organism initiates the desirable lactic acid fermentation in these products. It differs from other lactic acid species in that it can tolerate fairly high concentrations of salt and sugar (up to 50% sugar). *L. mesenteroides* initiates growth in vegetables more rapidly over a range of temperatures and salt concentrations than any other lactic acid bacteria. It produces carbon dioxide and acids which rapidly lower the pH and inhibit the development of undesirable micro-organisms. The carbon dioxide produced replaces the oxygen, making the environment anaerobic and suitable for the growth of subsequent species of lactobacillus. Removal of oxygen also helps to preserve the colour of vegetables and stabilises any ascorbic acid that is present.

Organisms from the gram positive *Propionibacteriaceae* family are responsible for the flavour and texture of some fermented foods, especially Swiss cheese, where they are responsible for the formation of 'eyes' or holes in the cheese. These bacteria break down lactic acid into acetic and propionic acids and carbon dioxide. Several other bacteria, for *instance Leuconostoc citrovorum L. Dextranicum, Streptococcus lactis, S. Cremis, & liquefaciens* and*Brevibacterium* species are important in the fermentation of dairy products. They are not discussed in detail in this manuscript.

Lactic acid fermentation

The lactic acid bacteria belong to two main groups – the homofermenters and the heterofermenters. The pathways of lactic acid production differ for the two. Homofermenters produce mainly lactic acid, via the glycolytic (Embden–Meyerhof) pathway). Heterofermenters produce lactic acid plus appreciable amounts of ethanol, acetate and carbon dioxide, via the 6-phosphoglucanate/phosphoketolase pathway. The glycolytic pathway is used by all lactic acid bacteria except leuconostocs, group III lactobacilli, oenococci and weissellas. Normal conditions required for this pathway are excess sugar and limited oxygen. Axelsson gives an in-depth account of the biochemical pathways for both homo- and hetero-fermenters.

Homolactic fermentation

The fermentation of 1 mole of glucose yields two moles of lactic acid;

$$\underset{\text{Glucose}}{C_6H_{12}O_6} \rightarrow \underset{\text{lactic acid}}{2\ CH_3CHOHCOOH}$$

Heterolactic fermentation

The fermentation of 1 mole of glucose yields 1 mole each of lactic acid, ethanol and carbon dioxide;

$$\underset{\text{Glucose}}{C_6H_{12}O_6} \rightarrow \underset{\text{lactic acid+}}{CH_3CHOHCOOH+} \quad \underset{\text{ethanol+}}{C_2H_5OH+} \quad \underset{\text{carbon dioxide}}{CO_2}$$

Acetic Acid Bacteria

A second group of bacteria of importance in food fermentations are the acetic acid producers from the *Acetobacter* species. *Acetobacter*are important in the production of vinegar (acetic acid) from fruit juices and alcohols. The same reaction also occurs in wines, oxygen permitting, where the *acetobacter* can cause undesirable changes – the oxidation of alcohol to acetic acid. This produces a vinegary off-taste in the wine.

The most desirable action of acetic acid bacteria is in the production of vinegar. The vinegar process is essentially a two stage process, where yeasts convert sugars into alcohol, followed by *acetobacter*, which oxidise alcohol to acetic acid.

Acetic acid fermentation

Acetobacter convert alcohol to acetic acid in the presence of excess oxygen.

Bacteria of Alkaline Fermentations

A third group of bacteria are those which bring about alkaline fermentations - the *Bacillus* species. Of note are *Bacillus subtilis, B. licheniformis* and *B.*

pumilius. Bacillus subtilis is the dominant species, causing the hydrolysis of protein to amino acids and peptides and releasing ammonia, which increases the alkalinity and makes the substrate unsuitable for the growth of spoilage organisms. Alkaline fermentations are more common with protein rich foods such as soybeans and other legumes, although there are a few examples utilising plant seeds. For example water melon seeds (*Ogiri* in Nigeria) and sesame seeds (*Ogiri-saro* in Sierra Leone) and others where coconut and leaf proteins are the substrates (Indonesian *semayi* and Sudanese *kawal* respectively).

Although the range of products of alkaline fermentation does not match those brought about by acid fermentations, they are important in that they provide protein rich, low cost condiments from leaves, seeds and beans, which contribute to the diet of millions of people in Africa and Asia. Steinkraus presents a comprehensive review of the acid, alkaline and alcoholic fermentations from around the world, which the reader is referred to for further information.

Conditions Required for Bacterial Fermentations

Micro-organisms vary in their optimal pH requirements for growth. Most bacteria favour conditions with a near neutral pH (7). The varied pH requirements of different groups of micro-organisms is used to good effect in fermented foods where successions of micro-organisms take over from each other as the pH of the environment changes. Certain bacteria are acid tolerant and will survive at reduced pH levels. Notable acid-tolerant bacteria include *the Lactobacillus* and *Streptococcus* species, which play a role in the fermentation of dairy and vegetable products.

Oxygen requirements vary from species to species. The lactic acid bacteria are described as microaerophilic as they carry out their reactions with very little oxygen. The acetic acid bacteria however, require oxygen to oxidise alcohol to acetic acid. In vinegar production, oxygen has to be made available for the production of acetic acid, whereas with wine it is essential to exclude oxygen to prevent oxidation of the alcohol and spoilage of the wine.

Temperature

Different bacteria can tolerate different temperatures, which provides enormous scope for a range of fermentations. While most bacteria have a temperature optimum of between 20 to 30°C, there are some (the thermophiles) which prefer higher temperatures (50 to 55°C) and those with colder temperature optima (15 to 20°C). Most lactic acid bacteria work best at temperatures of 18 to 22°C. The*Leuconostoc* species which initiate fermentation have an optimum of 18 to 22°C. Temperatures above 22°C, favour the *lactobacillus*species.

Salt concentration

Lactic acid bacteria tolerate high salt concentrations. The salt tolerance gives them an advantage over other less tolerant species and allows the lactic acid fermenters to begin metabolism, which produces acid, which further inhibits the growth of non-desirable organisms. Leuconostoc is noted for its high salt tolerance and for this reason, initiates the majority of lactic acid fermentations.

Water activity

In general, bacteria require a fairly high water activity (0.9 or higher) to survive. There are a few species which can tolerate water activities lower than this, but usually the yeasts and fungi will predominate on foods with a lower water activity.

Hydrogen ion concentration (pH)

The optimum pH for most bacteria is near the neutral point (pH 7.0). Certain bacteria are acid tolerant and will survive at reduced pH levels. Notable acid-tolerant bacteria include *the Lactobacillus* and *Streptococcus* species, which play a role in the fermentation of dairy and vegetable products.

Oxygen availability

Some of the fermentative bacteria are anaerobes, while others require oxygen for their metabolic activities. Some, lactobacilli in particular, are microaerophilic. That is they grow in the presence of reduced amounts of atmospheric oxygen. In aerobic fermentations, the amount of oxygen present is one of the limiting factors. It determines the type and amount of biological product obtained, the amount of substrate consumed and the energy released from the reaction. Acetobacter require oxygen for the oxidation of alcohol to acetic acid.

Nutrients

All bacteria require a source of nutrients for metabolism. The fermentative bacteria require carbohydrates – either simple sugars such as glucose and fructose or complex carbohydrates such as starch or cellulose. The energy requirements of micro-organisms are very high. Limiting the amount of substrate available can check their growth.

Principles of Lactic Acid Fermentation

Sauerkraut is one example of an acid fermentation of vegetables. The name sauerkraut literally translates as acid cabbage. The 'sauerkraut process' can be applied to any other suitable type of vegetable product. Because of the importance of this product in the German diet, the process has received substantial research in order to commercialise and standardise production. As a result, the process and the contributing micro-organisms are known intimately. Other less well known fermented fruits and vegetables have received less research attention, therefore little is known of the exact process. It is safe to assume however that the acid fermentation of vegetables is based on this process.

Lactic acid fermentations are carried out under three basic types of condition:– dry salted, brined and non-salted. Salting provides a suitable environment for lactic acid bacteria to grow which impart the acid flavour to the vegetable.

Dry salted fermented vegetables

With dry salting, the vegetable is treated with dry salt. The salt extracts the juice from the vegetable and creates the brine. The vegetable is prepared, washed in potable cold water and drained. For every 100 kg of vegetables 3 kg of salt is needed. The vegetables are placed in a layer of about 2.5cm depth in the fermenting container (a barrel or keg). Salt is sprinkled over the vegetables. Another layer of vegetables is added and more salt added. This is repeated until the container is three quarters full. A cloth is placed above the vegetables and a weight added to compress the vegetables and assist the formation of a brine which takes about 24 hours. As soon as the brine is formed, fermentation starts and bubbles of carbon dioxide begin to appear. Fermentation takes between one and four weeks depending on the ambient temperature. Fermentation is complete when no more bubbles appear, after which time the pickle can be packaged in a variety of mixtures. These can be vinegar and spices or oil and spices.

The 'sauerkraut' process

Lactic acid bacteria are the primary group of organisms involved in sauerkraut fermentation. They can be divided into three groups according to their types and end products:

Leuconostoc mesenteroides	an acid and gas producing coccus
Lactobacillus plantarum	bacilli that produce acid and a small amount of gas
Lactobacillus pentoaceticus	acid and gas producing bacilli

In addition to the desirable bacteria there are a range of undesirable micro-organisms present on cabbage which can interfere with the sauerkraut process if allowed to multiply unchecked. The quality of the final product depends largely on how well the undesirable organisms are controlled during the fermentation process. Some of the typical spoilage organisms utilise the protein as an energy source, producing unpleasant odours and flavours.

The fermentation process

Shredded cabbage or other suitable vegetables are placed in a jar and salt is added. Mechanical pressure is applied to the cabbage to expel the juice, which contains fermentable sugars and other nutrients suitable for microbial activity. The first micro-organisms to start acting are the gas-producing cocci *(L. Mesenteroides).* These microbes produce acids. When the acidity reaches 0.25 to 0.3% (calculated as lactic acid), these bacteria slow down and begin to die off, although their enzymes continue to function. The activity initiated by the *L. mesenteroides* is continued by the lactobacilli *(L. plantarum and L. Cucumeris)* until an acidity level of 1.5 to 2% is attained. The high salt concentration and low temperature inhibit these bacteria to some extent. Finally, *L. pentoaceticus* continues the fermentation, bringing the acidity to 2 to 2.5% thus completing the fermentation.

The end products of a normal kraut fermentation are lactic acid along with smaller amounts of acetic and propionic acids, a mixture of gases of which carbon dioxide is the principal gas, small amounts of alcohol and a mixture of aromatic esters. The acids, in combination with alcohol form esters, which contribute to the characteristic flavour of sauerkraut. The acidity helps to control the growth of spoilage and putrefactive organisms and contributes to the extended shelf life of the product. Changes in the sequence of desirable bacteria, or indeed the presence of undesirable bacteria, alter the taste and quality of the product.

Effects of temperature on sauerkraut process

The optimum temperature for sauerkraut fermentation is around 21°C. A variation of just a few degrees from this temperature alters the activity of the microbial process and affects the quality of the final product. Therefore, temperature control is one of the most important factors in the sauerkraut process. A temperature of 18° to 22° C is most desirable for initiating fermentation since this is the optimum temperature range for the growth and metabolism of *L. mesenteroides*. Temperatures above 22°C favour the growth of *Lactobacillus*species.

Effects of salt on the sauerkraut process

Salt plays an important role in initiating the sauerkraut process and affects the quality of the final product. The addition of too much salt may inhibit the desirable bacteria, although it may contribute to the firmness of the kraut. The principle function of salt is to withdraw juice from the cabbage (or other vegetable), thus making a more favourable environment for development of the desired bacteria.

Generally, salt is added to a final concentration of 2.0 to 2.5%. At this concentration, lactobacilli are slightly inhibited, but cocci are not affected. Unfortunately, this concentration of salt has a greater inhibitory effect against the desirable organisms than against those responsible for spoilage. The spoilage organisms can tolerate salt concentrations up to between 5 and 7%, therefore it is the acidic environment created by the lactobacilli that keep the spoilage bacteria at bay, rather than the addition of salt.

In the manufacture of sauerkraut, dry salt is added at the rate if 1 to 1.5 kg per 50kg cabbage (2 to 3%). The use of salt brines is not recommended in sauerkraut making, but is common in vegetables that have a low water content. It is essential to use pure salt since salts with added alkali may neutralise the acid.

Use of starter cultures

In order to produce sauerkraut of consistent quality, starter cultures (similar to those used in the dairy industry) have been recommended. Not only do starter cultures ensure consistency between batches, they speed up the fermentation process as there is no time lag while the relevant microflora colonise the sample. Because the starter cultures used are acidic, they also inhibit the undesirable micro-organisms. It is possible to add starters traditionally used for milk fermentation, such as *Streptococcus lactis*, without adverse effect on final quality. Because these organisms only survive for a short time (long enough to initiate the acidification process) in the kraut medium, they do not disturb the natural sequence of micro-organisms. On the other hand, if *Leuconostoc mesenteroides* is added in the early stages, it gives a good flavour to the final product, but alters the sequence of subsequent bacterial growth and results in a product that is incompletely fermented. If gas producing rods (for example *L pentoaceticus)* are added to the sauerkraut, this disturbs the balance between acetic and lactic acids - more acetic acid and less lactic acid are produced than normal - and the fermentation never reaches completion. If lactic acid, non-gas producing *rods (L. Cucumeris)* are used as a starter, again the kraut is not completely fermented and the resulting product is bitter and more susceptible to spoilage by yeasts.

It is possible to use the juice from a previous kraut fermentation as a starter culture for subsequent fermentations. The efficacy of using old juice depends largely on the types of organisms present in the juice and its acidity. If the starter juice has an acidity of 0.3% or more, it results in a poor quality kraut. This is because the cocci which would normally initiate fermentation are suppressed by the high acidity, leaving the bacilli with sole responsibility for fermentation. If the starter juice has an acidity of 0.25% or less, the kraut produced is normal, but there do not appear to be any beneficial effects of adding this juice. Often, the use of old juice produces a sauerkraut which has a softer texture than normal.

Spoilage and defects in the sauerkraut process

The majority of spoilage in sauerkraut is due to aerobic soil micro-organisms which break down the protein and produce undesirable flavour and texture changes. The growth of these aerobes can easily be inhibited by a normal fermentation.

Soft kraut can result from many conditions such as large amounts of air, poor salting procedure and varying temperatures. Whenever the normal sequence of bacterial growth is altered or disturbed, it usually results in a soft product. It is the lactobacilli, which seem to have a greater ability than the cocci to break down cabbage tissues, which are responsible for the softening. High temperatures and a reduced salt content favour the growth of lactobacilli, which are sensitive to higher concentrations of salt. The usual concentration of salt used in sauerkraut production slightly inhibits the lactobacilli, but has no effect on the cocci. If the salt content is too low initially, the lactobacilli grow too rapidly at the beginning and upset the normal sequence of fermentation.

Another problem encountered is the production of dark coloured sauerkraut. This is caused by spoilage organisms during the fermentation process. Several conditions favour the growth of spoilage organisms. For example, an uneven distribution of salt tends to inhibit the desirable organisms while at the same time allowing the undesirable salt tolerant organisms to flourish. An insufficient level of juice to cover the kraut during the fermentation allows undesirable aerobic bacteria and yeasts to grow on the surface of the kraut, causing off flavours and discoloration. If the fermentation temperature is too high, this also encourages the growth of undesirable microflora, which results in a darkened colour.

Pink kraut is a spoilage problem. It is caused by a group of yeasts which produce an intense red pigment in the juice and on the surface of the cabbage. It is caused by an uneven distribution of or an excessive concentration of salt, both of which allow the yeast to multiply. If conditions are optimal for normal fermentation, these spoilage yeasts are suppressed.

Brine salted fermented vegetables

Brine is used for vegetables which inherently contain less moisture. A brine solution is prepared by dissolving salt in water (a 15 to 20% salt solution). Fermentation takes place well in a brine of about 20 salometer. As a general guide, a fresh egg floats in a 10% brine solution. Properly brined vegetables will keep well in vinegar for a long time. The duration of brining is important for the overall keeping qualities. The vegetable is immersed in the brine and allowed to ferment. The strong brine solution draws sugar and water out of the vegetable, which decreases the salt concentration. It is crucial that the salt concentration does not fall below 12%, otherwise conditions do not allow for fermentation. To achieve this, extra salt is added periodically to the brine mixture.

Once the vegetables have been brined and the container sealed, there is a rapid development of micro-organisms in the brine. The natural controls which affect the microbial populations of the fermenting vegetables include the concentration of salt and temperature of the brine, the availability of fermentable materials and the numbers and types of micro-organisms present at the start of fermentation. The rapidity of the fermentation is correlated with the concentration of salt in the brine and its temperature.

Most vegetables can be fermented at 12.5° to 20° salometer salt. If so, the microbial sequence of lactic acid bacteria generally follows the classical sauerkraut fermentation described by Pederson. At higher salt levels of up to about 40° salometer, the sequence is skewed towards the development of a homofermentation, dominated by *Lactobacillus plantarum*. At the highest concentrations of salt (about 60° salometer) the lactic fermentation ceases to function and if any acid is detected during brine storage it is acetic acid, presumably produced by acid-forming yeasts which are still active at this concentration of salt.

Brine salted fermentation of vegetables (Pickles)

Pickled cucumbers are another fermented product that has been studied in detail and the process is known. The fermentation process is very similar to the sauerkraut process, only brine is used instead of dry salt. The washed cucumbers are placed in large tanks and salt brine (15 to 20%) is added. The cucumbers are submerged in the brine, ensuring that none float on the surface - this is essential to prevent spoilage. The strong brine draws the sugar and water out of the cucumbers, which simultaneously reduces the salinity of the solution. In order to maintain a salt solution so that fermentation can take place, more salt has to be added to the brine solution. If the concentration of salt falls below 12%, it will result in spoilage of the pickles through putrefaction and softening.

A few days after the cucumbers have been placed in the brine, the fermentation process begins. The process generates heat which causes the brine to boil rapidly. Acids are also produced as a result of the fermentation.

During fermentation, visible changes take place which are important in judging the progress of the process. The colour of the cucumber surface changes from bright green to a dark olive green as acids interact with the chlorophyll. The interior of the cucumber changes from white to a waxy translucent shade as air is forced out of the cells. The specific gravity of the cucumbers also increases as a result of the gradual absorption of salt and they begin to sink in the brine rather than floating on the surface.

Microbes involved in the fermentation process

As with the sauerkraut process, the gram positive coccus - *Leuconostoc mesenteroides* predominates in the first stages of pickle fermentation. This species is more resistant to temperature changes and tolerates higher salt concentration than the subsequent species. As fermentation proceeds and the acidity increases, lactobacilli start to take over from the cocci. The active stage of fermentation continues for between 10 to 30 days, depending upon the temperature of the fermentation. The optimum temperature for *L. Cucumeris*is 29 to 32°C. During the fermentative period, the acidity increases to about 2% and the strong acid producing types of bacteria reach their maximum growth. If sugar or acetic acid is added to the fermenting mixture during this time it increases the production of acid.

Problems in pickles

The production of excessive amounts of acid during the fermentation, results in shrivelling of the pickles, possibly due to over-activity of the *L. mesenteroides* species. If the brine is stirred, it may introduce air, which makes conditions more favourable for the growth of spoilage bacteria. In general, if the pickles are well covered with brine, the salt concentration is maintained and the temperature is at an optimum, it should be quite simple to produce good quality pickles.

Non salted, lactic acid fermented vegetables

Some vegetables are fermented by lactic acid bacteria, without the prior addition of salt or brine. Examples of non-salted products include*gundruk* (consumed in Nepal), *sinki* and other wilted fermented leaves. The detoxification of cassava through fermentation includes an acid fermentation, during which time the cyanogenic glycosides are hydrolysed to liberate the toxic cyanide gas.

The fermentation process relies on the rapid colonisation of the food by lactic acid producing bacteria, which lower the pH and make the environment unsuitable for the growth of spoilage organisms. Oxygen is also excluded as the *Lactobacilli* favour an anaerobic atmosphere. Restriction of oxygen ensures that yeasts do not grow.

For the production of *sinki*, fresh radish roots are harvested, washed and wilted by sun-drying for one to two days. They are then shredded, re-washed and packed tightly into an earthenware or glass jar, which is sealed and left to ferment. The optimum fermentation time is twelve days at 30°C. *Sinki* fermentation is initiated by *L. fermentum* and *L. brevis*, followed by *L. plantarum*. During fermentation the pH drops from 6.7 to 3.3. After fermentation, the radish substrate is sun-dried to a moisture level of about 21%. For consumption, *sinki* is rinsed in water for two minutes, squeezed to remove the excess water and fried with salt, tomato, onion and green chilli. The fried mixture is then boiled in rice water and served hot as soup along with the main meal.

Pit fermentations

South Pacific pit fermentations are an ancient method of preserving starchy vegetables without the addition of salt. The raw materials undergo an acid fermentation within the pit, to produce a paste with good keeping qualities. Pit fermentations are also used in other parts of the world – for example in Ethiopia, where the false banana (*Ensete ventricosum*) is fermented in a pit to produce a pulp known as*kocho*. Foods preserved in pits can last for years without deterioration, therefore pits provide a good, reliable cheap means of storage.

Root crops and bananas are peeled before being placed in the pit, while breadfruit are scraped and pierced. Food is left to ferment for three to six weeks, after which time it becomes soft, has a strong odour and a paste-like consistency. During fermentation, carbon dioxide builds up in the pit, creating an anaerobic atmosphere. As a result of bacterial activity, the temperature rises much higher than the ambient temperature. The pH of the fruit within the pit decreases from 6.7 to 3.7 within about four weeks. Inoculation of the fruit in the pit with lactic acid bacteria greatly speeds up the process. The fermented paste can be left in the pit and removed as required. Usually, it is removed and replaced with a second batch of fresh food to ferment. The fermented food is washed and fibrous material removed. It is then dried in the sun for several hours to remove the volatile odours, and pounded into a paste. Grated coconut or coconut cream and sugar may be added and the mixture is wrapped in banana leaves and either baked or boiled.

Principles of Acetic Acid Fermentation

The main desirable fermentation carried out by acetic acid bacteria is the production of vinegar. Vinegar, literally translated as sour wine, is one of the oldest products of fermentation used by man. It can be made from almost any fermentable carbohydrate source, for example fruits, vegetables, syrups and wine. Whatever the raw material used, the fermentation process follows a definite sequence.

The basic requirement for vinegar production is a raw material that will undergo an alcoholic fermentation. Apples, pears, grapes, honey, syrups, cereals, hydrolysed starches, beer and wine are all ideal substrates for the production of vinegar. The best raw materials are cider and wine, which are widely used in Europe and the United States. To produce a high quality product it is essential that the raw material is mature, clean and in good condition.

Microbes involved in the vinegar process

The production of vinegar depends on a mixed fermentation, which involves both yeasts and bacteria. The fermentation is usually initiated by yeasts which break down glucose into ethyl alcohol with the liberation of carbon dioxide gas.

The yeasts and bacteria exist together in a form known as commensalism. The acetobacter are dependent upon the yeasts to produce an easily oxidisable substance (ethyl alcohol). It is not possible to produce vinegar by the action of one type of micro-organism alone.

For a good fermentation, it is essential to have an alcohol concentration of 10 to 13%. If the alcohol content is much higher, the alcohol is incompletely oxidised to acetic acid. If it is lower than 13%, there is a loss of vinegar because the esters and acetic acid are oxidised. In addition to acetic acid, other organic acids are formed during the fermentation which become esterified and contribute to the characteristic odour, flavour and colour of the vinegar.

Acetaldehyde is an intermediate product in the transformation of the reducing sugar in fruit juice to acetic acid or vinegar. Oxygen is required for the conversion of acetaldehyde to acetic acid.

In general, the yield of acetic acid from glucose is approximately 60%. That is three parts of glucose yield two parts acetic acid.

Micro-organisms involved in the fermentation of vinegar

The organisms involved in vinegar production usually grow at the top of the substrate, forming a jelly like mass. This mass is known as 'mother of vinegar'. The mother is composed of both *acetobacter* and yeasts, which work together. The principal bacteria are*Acetobacter acetic A. Xylinum* and *A. Ascendens*. The main yeasts are *Saccharomyces ellipsoideus* and S *cerevisiae.* It is important to maintain an acidic environment to suppress the growth of undesirable organisms and to encourage the presence of desirable acetic acid producing bacteria. It is common practice to add 10 to 25% by volume of strong vinegar to the alcoholic substrate in order to attain a desirable fermentation.

The alcoholic fermentation of sugars should be completed before the solution is acidified because any remaining sugar will not be converted to alcohol after the acetic acid is added. Incomplete fermentation of the juice results in a "weak" product. The acetic acid strength of good vinegar should be approximately 6%.

Fermentation methods

Small scale production

Vinegar can be made at home at the small scale by introducing oxygen into barrels of wine or cider and allowing fermentation to occur spontaneously. This process is not very rigorously controlled and often results in a poor quality product.

The Orleans process

The Orleans process is one of the oldest and well known methods for the production of vinegar. It is a slow, continuous process, which originated in France. High grade vinegar is used as a starter culture, to which wine is added at weekly intervals. The vinegar is fermented in large (200 litre) capacity barrels. Approximately 65 to 70 litres of high grade vinegar is added to the barrel along with 15 litres of wine. After one week, a further 10 to 15 litres of wine are added and this is repeated at weekly intervals. After about four weeks, vinegar can be withdrawn from the barrel (10 to 15 litres per week) as more wine is added to replace the vinegar.

One of the problems encountered with this method is that of how to add more liquid to the barrel without disturbing the floating bacterial mat. This can be overcome by using a glass tube which reaches to the bottom of the barrel. Additional liquid is poured in through the tube and therefore does not disturb the bacteria. Wood shavings are sometimes added to the fermenting barrel to help support the bacterial mat.

Quick vinegar method

Because the Orleans process is slow, other methods have been adapted to try and speed up the process. The German method is one such method. It uses a generator, which is an upright tank filled with beechwood shavings and fitted with devices which allow the alcoholic solution to trickle down through the shavings in which the acetic acid bacteria are living. The tank is not allowed to fill as that would exclude oxygen which is necessary for the fermentation. Near the bottom of the generator are holes which allow air to be drawn in. the air rises through the generator and is used by the acetic acid bacteria to oxidise the alcohol. This oxidisation also releases considerable amounts of heat which must be controlled to avoid causing damage to the bacteria.

Problems in vinegar production

Many of the problems of vinegar production are concerned with the presence of nematodes, mites, flies and other insects. These pests can be controlled by adherence to good hygiene and pasteurisation of the vinegar. Problems associated with the fermentative process include the presence of a whitish film on the surface of the vinegar. This is sometimes called *Mycoderma vini* and is composed of yeast-like organisms, which grow aerobically and oxidise the carbon containing compounds to carbon dioxide and water. They also alter the flavour and alcohol content of the vinegar. This problem can however be controlled by adding one part vinegar to three parts of the alcoholic solution or by storing the alcoholic liquid in filled closed containers.

The Way Ahead

Fermentation is one of the oldest forms of food preservation technology in the world. Indigenous fermented foods such as bread, cheese and wine, have been prepared and consumed for thousands of years and are strongly linked to culture and tradition, especially in rural households and village communities. Fermented foods are popular throughout the world and in some regions make a significant contribution to the diet of millions of individuals. For instance Soy sauce is consumed throughout the world and is a fundamental ingredient in diets from Indonesia to Japan. Over one billion litres of soy sauce are produced each year in Japan alone. In Africa fermented cassava products (like *gari* and *fufu*) are a major component of the diet of more than 800 million people and in some areas these products constitute over 50% of the diet. Fermentation is a relatively efficient, low energy preservation process, which increases the shelf life and decreases the need for refrigeration or other forms of food preservation technology. It is therefore a highly appropriate technique for use in developing countries and remote areas where access to sophisticated equipment is limited. There is tremendous scope and potential for the use of micro-organisms towards meeting the growing world demand for food, through efficient utilisation of available natural food and feed stocks and the transformation of waste materials.

There is a danger that the introduction of 'western foods' with their glamorous image will displace these traditional fermented foods. Although fermentation of foods has been in use for thousands of years for the preservation and improvement of a range of foods, the microbial and enzymatic processes responsible for the transformations were, and still are, largely unknown. Because of the tremendously important role indigenous fermented fruits and vegetables play in food preservation and their potential to contribute to the growing food needs of the world, it is essential that the knowledge of their production is not lost. Moreover, it is essential to increase the knowledge and understanding of the methods of preparation, in order to improve the efficiency of fermentation, especially the traditional processes as the yields of traditional fermentation processes are often low and sometimes the products are unsafe. This section discusses potential areas for improvement of indigenous fermented fruit and vegetable products. It can be divided into four main areas:

— Improve the understanding of the fermented products
— Refine the processes
— Disseminate the improvements
— Create a supportive policy environment.

Improving the Understanding of Fermented Products

For fermented products such as cheese, bread, beer and wine, which are produced on a commercial scale, a good understanding of the microbial processes has been developed. However with many of the fermented products in Africa, Asia and Latin America, knowledge of the processes involved is poor. It is likely that the basic principles apply across the board, but production conditions vary enormously from region to region, giving rise to numerous variations of the basic fermented product. It is not the intention or the desire to standardise the process and thereby lose this huge diversity, rather it is to harness the tremendous potential these methods have to contribute to increasing not only the quantity, but quality of food available to the worlds population.

There are two main reasons for gaining a better understanding of indigenous fermented products:

— Documenting the traditional knowledge to ensure that the huge diversity is not lost

— Developing a scientific understanding of the microbial processes, with a view to improving the efficiency of the process

Documenting the Traditional Knowledge

Fermentation is one of the oldest food processing technologies in the world. The knowledge of how to make these products has often been passed down from parent to child (usually mother to daughter) and belongs to that undervalued body of "indigenous knowledge". Most of this knowledge has not been documented and is in danger of being lost as technologies evolve and families move away from traditional food preservation practices.

Several research institutes and scientists in Africa, Asia and Latin America are recording information on traditional fermented foods. Dr Karki in Nepal and Hamid Dirar have been extensively quoted in this book and their work on fermented products in Nepal and Sudan is of great importance.

Intermediate Technology has been collecting information about traditional food products from Africa, Asia and Latin America. The first volume of products was published in 1997. This included the following fermented food products: *kenkey*, *ogi*, *injera*, fermented sweet bread, sorghum beer, palm wine, banana beer, pineapple peel vinegar, lime pickle, tamarind pickle, mango pickle, *gari*, vegetable pickle, coconut toddy, yoghurt, *ayib* and *dawdawa*. Volume 2 will be published in 1999 and regional publications on the traditional food products of Bangladesh, Southern Africa and the Andes are planned.

The Special Programme on Biotechnology and Development Cooperation for the Netherlands Government was established in 1992 to improve the access of developing countries to biotechnological expertise and innovation with a focus on using biotechnology for the benefit of small-scale farmers and producers. Part of the programme involved a competition to identify farmers' existing biotechnology practises. This programme resulted in the collection of valuable information about traditional fermented food products.

Finally, the Food and Agriculture Organisation of the United Nations sees the value in collecting and preserving this source of knowledge. This book is an example of their work.

Developing a Scientific Understanding of the Microbial Processes

Most traditional fermented products are made by natural fermentations carried out in a non-sterile environment. The specific environmental conditions cause a gradual selection of micro-organisms responsible for the desired final product. This is appropriate for small-scale production for home consumption. However the method is difficult to control and there are risks of accompanying micro-flora causing spoilage and unsafe products.

If the processes are to be refined, with a view to production on a larger scale, it is essential to have a scientific understanding of the fermentation processes. This can be developed by:

- The isolation and characterisation of the essential micro-organisms involved;
- The determination of the role of external factors in fermentations and the effects of these on the metabolism of micro-organisms
- The investigation of the effects of pre-treatments of raw materials on the fermentation process
- The identification of the options for further processing and how these affect the taste and texture of the product.

This research is capital intensive and usually requires scarce foreign exchange. It requires the use of sophisticated equipment and reagents backed with a consistent energy and water supply which are not always available in developing countries. To meet the current and future challenges in developing countries, it is important that these countries develop the capabilities to benefit from improvements in fermentation methodologies.

Biotechnologies need to be developed which are affordable by the poor, since it is they who are likely to benefit most by improvements to the traditional processes.

Refining the Process

The art of traditional processes needs to be refined to incorporate objective methods of process control and to standardise quality of the final product without losing their desirable attributes such as improved keeping quality, taste and nutritional qualities.Science based fermentation research has often focused mainly on improving the metabolic properties of the micro-organisms used as a starter culture, using the techniques of selection, mutation and genetic modification. To achieve this a full understanding of the fermentation process is needed. The properties of the starter culture must be known and culture conditions must be controllable. This means that science based fermentation research currently has little to offer traditional food processing.

Process control

Once the details of the fermentation process and the microbes involved are known and understood, it is possible to begin to refine and improve the process. The commercially produced products, such as bread, wine, soy sauce and pickles are all examples of processes which have been studied and optimised. Recently traditional fermented products in Africa have been industrialised. In Nigeria, Dadwa (a fermented legume product) is now made by Cadburys and in Zimbabwe, traditional fermented milk is made industrially and sold as "Lacto".

The areas where the efficiency and yield of food fermentation processes can be increased are:

— The selection or development of more productive microbial strains

— The control and manipulation of culture conditions

— The improvement of product purification and concentration.

Techniques to control the fermentation processes could include the development of pure starter cultures. Developing these by laboratory selection or genetic engineering is not viable. A more feasible approach would be to exploit the ecological principle of inoculum enrichment by natural selection. Another approach to stabilise fermentation under nonsterile conditions is the use of multi-strain dehydrated starters which can be stored at ambient temperature. These are already used for the manufacture of tempeh.

Quality control

The aim of quality control is to ensure that every batch of food produced has a satisfactory and uniform quality. This does not necessarily mean that it is the highest quality possible but that it reaches the standard the customers are willing to pay for.

Inadequate quality control can have an adverse effect on local demand for the product. This is particularly a problem for small-scale traditional production. In modern industrial applications, the fermentation equipment and processes are controlled using expensive technology, resulting in a consistent product of a known quality. Traditional practises take place in a less predictable environment. This can result in mistakes including sour beer and mouldy pickles.

It is often felt that traditional products made at the small scale are unhygienic and unsafe. This is sometimes true. However the case is often overstated. Many fermented foods are inherently safe due to low moisture contents or high acidity. Lime pickle from India and Gundruk from Nepal are good examples. Several of the steps in traditional processing are designed to reduce contamination. These include boiling, adding salt and sun drying.

Quality control procedures are essential for the production of safe products and contribute to the success of small food processing businesses. Appropriate quality control procedures need to be developed and implemented. These procedures need to be developed with the processors who must understand and apply them. The quality of food is highly subjective. What is acceptable to one customer is not acceptable to another. It is important to carry out participative research to identify ways to improve the quality control procedures for fermented food products.

The sort of areas that should be investigated include:

— Selecting good quality raw materials
— Processing under correct conditions
— Ensuring high standards of personal hygiene by the food processors
— Ensuring the processing area is sufficiently clean
— Using correct packaging

Disseminating Improvements

Documentation of the traditional methods of food fermentation and research to identify improved methods of production are meaningless if the results are not disseminated to those who are likely to put them into practice.

There is a danger of mystifying the fermentation process by enrobing it in scientific theory. What was once a simple process carried out by any family member, in the confines of the household, using locally available equipment and materials, could become a process to be feared. It is important to be realistic and to ensure that the improvements recommended are ones which can easily be put into practice. The aim is not to deter the production of fermented foods at the small scale, but to encourage their production and consumption on a larger scale.

Fermented foods often have a stigma attached to them – they are considered as poor mans food. As soon as a family can afford to buy processed foods, they move away from carrying out home fermentation. This is a pity, because as we have seen earlier, fermented food products have many nutritional advantages which surpass western-style fast foods and processed foods. Where cultural values attached to the fermented food are strong, it is unlikely that there will be an image problem. For example, kimchi is considered part of the national heritage in Korea. It is a vital ingredient of all meals and as such is a highly valued food. This is reflected in the amount of research carried out on the product and the detailed understanding of the process, which already exists.

It is not difficult to gain access to village people, both to collect the traditional information and to disseminate improved practices. Numerous organisations are involved in field projects. They can be used to organise training sessions and group meetings for the dissemination of new methods, for example, for the use of pure starter cultures.

Creating a Supportive Environment for Production of Fermented Food Products

Developing countries need to build their resources of trained, knowledgeable individuals, who are able to apply the basic microbiological principles to the production of fermented foods. Extra support should be made available to train professionals in this discipline.

Fermented foods should be recognised as part of each countries heritage and culture and efforts made to preserve the methods of production. A recognised body (government or non-government) should take the responsibility for the collection of details and the promotion of fermented food products. Consumers need to be made aware of the numerous benefits of fermented foods and their prejudices against fermented foods, especially those traditionally produced at the home scale, dispelled. Many traditional fermented foods are produced from minor or wild fruits and vegetables, many of which are being lost through deforestation and loss of biodiversity.

References

Aidoo, K.E., 1986, *Lesser-known fermented plant foods*, Tropical Science, 1986, 26, 249-258.

Fellows, P., 1997, *Traditional Foods*, Intermediate Technology Publications, UK.

Kordylas, J.M., (1990), *Processing and Preservation of Tropical and Subtropical Foods,* Macmillan, UK.

Pedersen, C.S., 1979, *Microbiology of Food Fermentations*, AVI Publishers, USA.

Steinkraus K H, 1996. *Handbook of Indigenous Fermented Foods*. Marcel Decker Inc, New York.

Ottogalli, G. and Galli, A., 1997, *Fermented Foods in the Past and in the Future*, Annali di Microbiologia ed Enzimologia, Italy.

4

Microbial Food Spoilage

Food spoilage is a metabolic process that causes foods to be undesirable or unacceptable for human consumption due to changes in sensory characteristics. Spoiled foods may be safe to eat, i.e. they may not cause illness because there are no pathogens or toxins present, but changes in texture, smell, taste, or appearance cause them to be rejected. Some ecologists have suggested these noxious smells are produced by microbes to repulse large animals, thereby keeping the food resource for themselves ! Food loss, from farm to fork, causes considerable environmental and economic effects.

Shelf life of a food is the time during which it remains stable and retains its desired qualities. Some spoilage is inevitable, and a variety of factors cause deterioration of foods:

— endogenous enzymes in plants oxidizing phenolic compounds (browning) or degrading pectins (softening);

— insects infesting foods and rodents chewing on foods;

— parasites, when visible for example in meat or fish, rendering food undesirable;

— microbes (bacteria, molds, yeasts) growing on and metabolizing foods;

— light causing degradation of pigments, fats, and proteins (off-flavors and odors) or stimulating pigment production (greening of potatoes);

— temperature; both excessive heat and freezing physically affecting texture of foods and breaking emulsions;

— air, particularly oxygen, oxidizing lipids producing strong off-odors and flavors;

— moisture: too little causing cracking, crumbling, or crystallization whereas excess causes sogginess, stickiness, or lumping.

These factors are interrelated, as certain temperatures and oxygen and moisture levels increase the activities of endogenous enzymes and of microbes. Rodent and insect damage may provide an entry point for microbial growth.

Food spoilage is a broad topic that cannot be completely addressed in one review article. This chapter will emphasize spoilage caused by microorganisms and will consider spoilage of foods that people purchase or consume. For example, spoilage of bread will be considered but not deterioration of wheat plants in the fields or wheat grains in storage. Non-microbial spoilage such as loss of water (shriveling of greens or carrots) or changes induced by degradative enzymes in plants (yellowing of broccoli) will not be covered. Pathogenic organisms, e.g. *Listeria* that cause human illness, will not be considered even though they may also cause some spoilage.

Detection of Spoilage

Spoilage is manifested by a variety of sensory cues such as off-colors, off-odors, softening of vegetables and fruits, and slime. However, even before it becomes obvious, microbes have begun the process of breaking down food molecules for their own metabolic needs. Sugars and easily digested carbohydrates are used first, plant pectins are degraded. Then proteins are attacked, producing volatile compounds with characteristic smells such as ammonia, amines, and sulfides. These odors start to develop in meat when there are about 10^7 cfu of bacteria/cm^2 of meat surface and are usually recognizable at populations of 10^8 cfu/cm^2.

Early detection of spoilage would be advantageous in reducing food loss because there may be interventions that could halt or delay deterioration, and on the other hand food that had reached the end of its designated shelf life but was not spoiled could still be used. Numerous methods for detection of spoilage have been devised with the goals of determining concentrations of spoilage microbes or volatile compounds produced by these microbes. However, many of these methods are considered inadequate because they are time-consuming, labor-intensive, and/or do not reliably give consistent results. Some representative papers using different methods are described below.

Traditional methods of estimating bacterial populations do not provide results quickly enough to allow for interventions. Microbial population levels can be measured by real time PCR in liquids, such as *Gluconobacter* in an electrolyte replacement drink and yeasts in wine and fruit juices. Other methods for detecting bacteria include ATP bioluminescence and electrical impedance assays but some food matrices may contain interfering substances.

Detection of volatile compounds produced by spoilage bacteria can be a less invasive and more rapid means for monitoring spoilage. Biogenic amines (putrescine, cadaverine, histamine, and tyramine) are commonly produced during spoilage of high protein foods and can attain levels that cause illness, particularly in spoiled fish. HPLC methods have been used to quantitate different amines in fish, chicken, and cheese. Combined concentrations of these amines are expressed as a bio-genic amine index that is related to the extent of food spoilage and to the concentrations of spoilage organisms.

Electronic noses were first developed about twenty years ago and have undergone many refinements since. They consist of a set of sensors that react with different volatile chemicals and produce an electrical signal. An odor profile can be analyzed by using pattern recognition files. Such systems have been used to detect spoilage in beef, bakery products, fish, and milk. Other techniques being developed to detect microbes or chemicals associated with spoilage include: (a) FT-IR (Fourier Transform-Infrared Spectroscopy) used with beef and apple juice ; (b) visible and short-wavelength near-infrared spectroscopy to detect microbial load in chicken by diffuse reflectance ; (c) ion mobility spectrometry for detecting trimethyl-amine in meat ; and (d) gas chromatography-mass spectrometry for analyses of fish. It is expected that some advances in nanotechnology will improve the portability, sensitivity, and speed of detection systems.

Spoilage Organisms

Chemical reactions that cause offensive sensory changes in foods are mediated by a variety of microbes that use food as a carbon and energy source. These organisms include prokaryotes (bacteria), single-celled organisms lacking defined nuclei and other organelles, and eukaryotes, single-celled (yeasts) and multicellular (molds) organisms with nuclei and other organelles. Some microbes are commonly found in many types of spoiled foods while others are more selective in the foods they consume; multiple species are often identified in a single spoiled food item but there may be one species (a specific spoilage organism, SSO) primarily responsible for production of the compounds causing off-odors and flavors. Within a spoiling food, there is often a succession of different populations that rise and fall as different nutrients become available or are exhausted. Some microbes, such as lactic acid bacteria and molds, secrete compounds that inhibit competitors.

Spoilage microbes are often common inhabitants of soil, water, or the intestinal tracts of animals and may be dispersed through the air and water and by the activities of small animals, particularly insects. It should be noted that with the development of new molecular typing methods, the scientific names of some spoilage organisms, particularly the bacteria, have changed in recent years and some older names are no longer in use. Many insects and small mammals also cause deterioration of food but these will not be considered here.

Yeasts

Yeasts are a subset of a large group of organisms called fungi that also includes molds and mushrooms. They are generally single-celled organisms that are adapted for life in specialized, usually liquid, environments and, unlike some molds and mushrooms, do not produce toxic secondary metabolites. Yeasts can grow with or without oxygen (facultative) and are well known for their beneficial fermentations that produce bread and alcoholic drinks. They often colonize foods with a high sugar or salt content and contribute to spoilage of maple syrup, pickles, and sauerkraut. Fruits and juices with a low pH are another target, and there are some yeasts that grow on the surfaces of meat and cheese. There are four main groups of spoilage yeasts:

Zygosaccharomyces and related genera tolerate high sugar and high salt concentrations and are the usual spoilage organisms in foods such as honey, dried fruit, jams and soy sauce. They usually grow slowly, producing off-odors and flavors and carbon dioxide that may cause food containers to swell and burst. *Debaryomyces hansenii* can grow at salt concentrations as high as 24%, accounting for its frequent isolation from salt brines used for cured meats, cheeses, and olives. This group also includes the most important spoilage organisms in salad dressings.

Saccharomyces spp. are best known for their role in production of bread and wine but some strains also spoil wines and other alcoholic beverages by producing gassiness, turbidity and off-flavors associated with hydrogen sulfide and acetic acid. Some species grow on fruits, including yogurt containing fruit, and some are resistant to heat processing.

Candida and related genera are a heterogeneous group of yeasts, some of which also cause human infections. They are involved in spoilage of fruits, some vegetables and dairy products.

Dekkera/Brettanomyces are principally involved in spoilage of fermented foods, including alcoholic beverages and some dairy products. They can produce volatile phenolic compounds responsible for off-flavors.

Molds

Molds are filamentous fungi that do not produce large fruiting bodies like mushrooms. Molds are very important for recycling dead plant and animal remains in nature but also attack a wide variety of foods and other materials useful to humans. They are well adapted for growth on and through solid substrates, generally produce airborne spores, and require oxygen for their metabolic processes. Most molds grow at a pH range of 3 to 8 and some can grow at very low water activity levels (0.7–0.8) on dried foods. Spores can tolerate harsh environmental conditions but most are sensitive to heat treatment. An exception is *Byssochlammys,* whose spores have a D value of 1–12 minutes at 90°C. Different mold species have different optimal growth temperatures, with some able to grow in refrigerators. They have a diverse secondary metabolism producing a number of toxic and carcinogenic mycotoxins. Some spoilage molds are toxigenic while others are not. Spoilage molds can be categorized into four main groups:

Zygomycetes are considered relatively primitive fungi but are widespread in nature, growing rapidly on simple carbon sources in soil and plant debris, and their spores are commonly present in indoor air. Generally they require high water activities for growth and are notorious for causing rots in a variety of stored fruits and vegetables, including strawberries and sweet potatoes. Some common bread molds also are zygomycetes. Some zygomycetes are also utilized for production of fermented soy products, enzymes, and organic chemicals. The most common spoilage species are *Mucor* and *Rhizopus*. Zygomycetes are not known for producing mycotoxins but there are some reports of toxic compounds produced by a few species.

Penicillium and related genera are present in soils and plant debris from both tropical and Antarctic conditions but tend to dominate spoilage in temperate regions. They are distinguished by their reproductive structures that produce chains of conidia. Although they can be useful to humans in producing antibiotics and blue cheese, many species are important spoilage organisms, and some produce potent mycotoxins (patulin, ochratoxin, citreoviridin, peni-trem). *Penicillium* spp. cause visible rots on citrus, pear, and apple fruits and cause enormous losses in these crops. They also spoil other fruits and vegetables, including cereals. Some species can attack refrigerated and processed foods such as jams and margarine. A related genus, *Byssochlamys*, is the most important organism causing spoilage of pasteurized juices because of the high heat resistance of its spores.

Aspergillus and related molds generally grow faster and are more resistant to high temperatures and low water activity than *Penicillium* spp. and tend to dominate spoilage in warmer climates. Many asper-gilli produce mycotoxins: aflatoxins, ochratoxin, ter-ritrems, cyclopiazonic acid. Aspergilli spoil a wide variety of food and non-food items (paper, leather, etc.) but are probably best known for spoilage of grains, dried beans, peanuts, tree nuts, and some spices.

Other molds, belonging to several genera, have been isolated from spoiled food. These generally are not major causes of spoilage but can be a problem for some foods. *Fusarium* spp. cause plant diseases and produce several important mycotoxins but are not important spoilage organisms. However, their myco-toxins may be present in harvested grains and pose a health risk.

Bacteria

Spore-forming bacteria are usually associated with spoilage of heat-treated foods because their spores can survive high processing temperatures. These Gram-positive bacteria may be strict anaerobes or facultative (capable of growth with or without oxygen). Some spore-formers are thermophilic, preferring growth at high temperatures (as high as 55°C). Some anaerobic thermophiles produce hydrogen sul-fide (*Desulfotomaculum*) and others produce hydrogen and carbon dioxide (*Thermoanaerobacterium*) during growth on canned/hermetically sealed foods kept at high temperatures, for example, soups sold in vending machines. Other thermophiles (*Bacillus* and *Geobacillus* spp.) cause a flat sour spoilage of high or low pH canned foods with little or no gas production, and one species causes ropiness in bread held at high ambient temperatures. Mesophilic anaerobes, growing at ambient temperatures, cause several types of spoilage of vegetables (Bacillus spp.); putrefaction of canned products, early blowing of cheeses, and butyric acid production in canned vegetables and fruits (*Clostridium* spp.); and "medicinal" flavors in canned low-acid foods (*Alicyclobacillus*). Psychrotolerant sporeformers produce gas and sickly odors in chilled meats and brine-cured hams (*Clos-tridium* spp.) while others produce off-odors and gas in vacuum-packed, chilled foods and milk (*Bacillus* spp.).

Lactic acid bacteria (LAB) are a group of Gram-positive bacteria, including species of *Lacto-bacillus, Pediococcus, Leuconostoc* and *Oenococcus*, some of which are useful in producing fermented foods such as yogurt and pickles. However, under low oxygen, low temperature, and acidic conditions, these bacteria become the predominant spoilage organisms on a variety of foods. Undesirable changes caused by LAB include greening of meat and gas formation in cheeses (blowing), pickles (bloater damage), and canned or packaged meat and vegetables. Off-flavors described as mousy, cheesy, malty, acidic, buttery or liver-like may be detected in wine, meats, milk, or juices spoiled by these bacteria. LAB may also produce large amounts of an exopolysac-charide that causes slime on meats and ropy spoilage in some beverages.

Pseudomonas and related genera are aerobic, Gram-negative soil bacteria, some of which can degrade a wide variety of unusual compounds. They generally require a high water activity for growth (0.95 or higher) and are inhibited by pH values less than 5.4. Some species grow at refrigeration temperatures (psychrophilic) while other are adapted for growth at warmer, ambient temperatures. Four species of *Pseudomonas* (*P. fluorescens, P. fragi, P. lundensis,* and *P. viridiflava*), *Shewanella putrefa-ciens*, and *Xanthomonas campestris* are the main food spoilage organisms in this group. Soft rots of plant-derived foods occur when pectins that hold adjacent plant cells together are degraded by pectic lyase enzymes secreted by *X. campestris, P. fluorescens* and *P. viridiflava.* These two species of *Pseudo-monas* comprise up to 40% of the naturally occurring bacteria on the surface of fruits and vegetables and cause nearly half of post-harvest rot of fresh produce stored at cold temperatures. *P. fluorescens, P. fragi, P. lundensis,* and *S. putrefaciens* cause spoilage of animal-derived foods (meat, fish, milk) by secreting lipases and proteases that cause formation of sulfides and trimethylamine (off-odors) and by forming biofilms (slime) on surfaces. Some strains are adapted for growth at cold temperatures and spoil these foods in the refrigerator.

Enterobacteriaceae are Gram-negative, facultatively anaerobic bacteria that include a number of human pathogens *(Salmonella, E coli, Shigella, Yersinia)* and also a large number of spoilage organisms. These bacteria are widespread in nature in soil, on plant surfaces and in digestive tracts of animals and are therefore present in many foods. *Erwinia carotovora* is one of the most important bacteria causing soft rot of vegetables in the field or stored at ambient temperatures. Biogenic amines are produced in meat and fish by several members of this group while others produce off-odors or colors in beer *(Obesumbacterium),* bacon and other cured meats *(Proteus, Serratia),* cheeses (several genera), cole slaw *(Klebsiella),* and shell eggs *(Proteus, Entero-bacter, Serratia).* Temperature, salt concentration, and pH are the most important factors determining which, if any, of these microbes spoil foods.

Many Gram-negative bacteria, including pseudo-monads and enterobacteriaceae, secrete acyl homo-serine lactones (AHLs) to regulate the expression of certain genes, such as virulence factors, as a function of cell density. These AHL quorum-sensing signals may regulate proteolytic enzyme production and iron chelation during spoilage of some foods although the role of these signals in other spoilage systems is not clear.

Other bacteria are associated with spoilage of chilled, high protein foods such as meat, fish, and dairy products. They may not be the predominant spoilage organisms but contribute to the breakdown of food components and may produce off-odors. Most species are aerobic although some grow at low oxygen levels and may survive vacuum packaging, and one *(Brochothrix)* is a facultative anaerobe. Some examples include:

— *Acinetobacter* and *Psychrobacter,* which are predominant bacteria on poultry carcasses on the processing line and have been isolated from a variety of spoiled meat and fish. *Acinetobacter* grows at a pH as low as 3.3 and has been detected in spoiled soft drinks. These two genera do not produce extracellular lipases, hydrogen sulfide, or trimethylamine (fishy odor) and so are considered to have a low spoilage potential.

— *Alcaligenes* is a potential contaminant of dairy products and meat and has been isolated from rancid butter and milk with an off-odor. These bacteria occur naturally in the digestive tract of some animals and also in soil and water.

— *Flavobacterium* is found widely in the environment and in chilled foods, particularly dairy products, fish, and meat. It uses both lipases and proteases to produce disagreeable odors in butter, margarine, cheese, cream, and other products with dairy ingredients.

— *Moraxella* and *Photobacterium* are important constituents of the microflora on the surface of fish. *Photobacterium* can grow and produce trimethylamine in ice-stored, vacuum-packaged fish.

— *Brochothrix* has been isolated from meat, fish, dairy products and frozen vegetables. During spoilage, it produces odors described as sour, musty, and sweaty.

Modelling Spoilage

Several physical factors determine whether spoilage microbes will be successful in utilizing the nutrients in a food. These include water activity and types of solutes, pH, temperature (storage and processing), oxygen and carbon dioxide levels, solid or liquid state of food, available nutrients, and preservatives. Models for microbial spoilage of different foods and for spoilage by specific organisms examine the effects of these factors—singly and in combination— to predict the initiation and course of the spoilage process. These models are based on and validated with actual experimental data and can provide useful information for product development and modification, shelf-life estimates, processing requirements, and quality assurance programs.

Food spoilage is a complex process involving a variety of organisms, food preservatives and additives, and food matrices in addition to temperature, pH, and water activity, the most important determinants of microbial growth. Depending on their objective, models are constructed to focus on probability of growth/no growth, time required to initiate growth, growth rate, or survival of spoilage organisms under a particular set of parameters. Inactivation and destruction of microbes exposed to different preservatives or preservation techniques can also be modeled. However, models cannot incorporate every factor that may affect the spoilage process and processors should validate models for their own products to account for different variables. Examples of some recent models are described below:

Growth/no growth was modeled for yeast in a simulated fruit-based or alcoholic food or drink as a function of temperature, pH, and sucrose, sorbitol, and alcohol levels. The model could predict the likelihood of growth and the time to growth.

Enzyme activity of lipases from five spoilage bacteria were modeled as a function of temperature, pH, and water activity.

Inactivation/destruction of Saccacharomyces cerevisiae in orange juice by high intensity pulsed electric fields was modeled as a function of field strength and time of exposure.

Growth rate models:

— *Pseudomonas fluoresces* at 7°C as a function of different levels of oxygen (5-75%), carbon dioxide (0-15%) and nitrogen (10-80%) validated on cut lettuce.

— *Leuconostoc mesenteroides* under aerobic and anaerobic conditions as a function of temperature, pH, and sodium chloride and sodium nitrite levels.

— *Alicyclobacillus* in orange juice as a function of temperature, pH, soluble solids concentration (°Brix) and nisin levels.

— *Brochothrix* as a function of temperature, pH, and water activity.

— Cocktail of *Pseudomonas, Shewanella,* and *Acinetobacter* as a function of temperature, pH, and water activity.

— Cocktail of several strains of Enterobacteri-aceae as a function of temperature, pH, and water activity.

— Spoilage bacteria on ground meat as a function of temperature and pH.

— *Aspergillus niger,* a mold, as a function of temperature and water activity.

— *Penicillium italicum,* a mold, as a function of temperature, water activity and solute levels (glycerol, sorbitol, glucose, NaCl).

— Molds *(Eurotium, Aspergillus, Penicillium)* on sponge cake as a function of pH, water activity, and carbon dioxide levels and in fermented bakery product analogues as a function of pH, water activity, and potassium sorbate levels.

— Yeast, *Pichia anomala,* associated with olive fermentation, as a function of temperature, pH, and NaCl.

— Yeast, *Zygosaccharomyces bailii,* as a function of pH and benzoic acid levels.

— Cocktail of three spoilage yeasts *(Yarrowia, Pichia, and Zygosaccharomyces)* as a function of temperature, pH, and water activity.

Factors Affecting Food Spoilage and Shelf Life

Foods by their nature are rich in carbohydrates, proteins and lipids that microbes as well as humans find very nutritious. Living plants and animals have structural and chemical defenses to prevent microbial colonization, but once they are dead or in a dormant state these systems deteriorate and become less effective. Many different microbes may potentially be able to use the nutrients in a food but some species have a competitive advantage under certain conditions. Food processors should note that certain spoilage organisms may not grow on particular foods because some nutrient is missing. If the food product is reformulated, then a new ingredient may allow growth of a previously unimportant microbe. Different food categories present different challenges for inhibition of spoilage organisms.

Dairy Products

Milk is an excellent medium for growth for a variety of bacteria. Spoilage bacteria may originate on the farm from the environment or milking equipment or in processing plants from equipment, employees, or the air. LAB are usually the predominant microbes in raw milk and proliferate if milk is not cooled adequately. When populations reach about 10^6 cfu/ml, off-flavors develop in milk due to production of lactic acid and other compounds. Refrigeration suppresses growth of LAB and within one day psychro-philic bacteria (*Pseudomonas, Enterobacter, Alcali-genes* and some spore-formers) grow and can eventually produce rancid odors through the action of lipases and bitter peptides from protease action. Pasteurization kills the psychrophiles and mesophilic bacteria (LAB), but heat-tolerant species (*Alcali-genes*, *Microbacterium,* and the sporeformers *Bacillus* and *Clostridium*) survive and may later cause spoilage in milk or other dairy products. Immediately following pasteurization, bacterial counts are usually <1000 cfu/ml. However, post-pasteurization contamination of milk, particularly with *Pseudomonas* and some Gram-positive psychrophiles does occur.

Spoilage problems in cheese can sometimes be traced to low quality milk but may also result from unhygienic conditions in the processing plant. Hard and semi-hard cheeses have a low moisture content (<50%) and a pH ~5.0, which limits the growth of some microbes. Some coliforms and *Clostridium* spp. that cause late gas blowing can grow under these conditions as can several species of molds. Other psychrotrophs produce biogenic amines, particularly tyramine, during storage of cheese. Soft cheeses with a higher pH of 5.0–6.5 and a moisture content of 50–80% may be spoiled by *Pseudomonas, Alcali-genes,* and *Flavobacterium. Clostridium sporogenes* has been found in spoiled processed cheese, where it produces gas holes and off-flavors.

Yeasts and molds are the main spoilage organisms found in cultured milks (yogurt, sour cream and buttermilk) because the higher acidity in these products inhibits many bacteria. *Pseudomonas*, yeasts and molds can spoil butter and "light" butters. Since the light butters have a higher moisture content than butter, they can support more microbial growth. Cream may become rancid when populations of *Pseudomonas* and *Enterobacter* proliferate.

Cereal and Bakery Products

Cereal grains are exposed to a variety of bacteria, molds and yeasts during growth, harvesting, drying and storage. Molds are the most important contaminants because of the low moisture levels in grains, but molds do require some moisture so efficient drying and good storage facilities are necessary to prevent their growth. Microbial populations decrease during milling and storage of grain. Molds cause spoilage by altering the appearance of grains and flours, and some species also synthesize toxic secondary metabolites called mycotoxins.

Molds are also the primary spoilage organisms in baked goods, with *Aspergillus*, *Penicillium*, and *Eurotium* being the most commonly isolated genera. *Penicillium* tends to be the more important in sourdough breads and in breads stored at cooler temperatures. Freshly baked breads do not contain viable molds but soon become contaminated upon exposure to air and surfaces. *Bacillus* spores are very heat resistant and can survive baking in the interior of bread loaves and then germinate and start growing as the bread cools. Some strains cause a defect called ropiness, a soft sticky texture caused by starch degradation and slimy exopolysaccharides often accompanied by a fruity odor. Yeasts may also be involved in spoilage of some breads and fruitcakes, causing a chalky appearance on surfaces and off-odors.

High sugar content and low water activity of cakes also favors molds over other spoilage microbes but some species of yeasts and bacteria (*Bacillus* and *Pseudomonas*) may also attack cakes. Bakery products containing cream, custard or fruit fillings are targets of additional spoilage organisms.

Vegetables

Vegetables are another tempting source of nutrients for spoilage organisms because of their near neutral pH and high water activity. Although vegetables are exposed to a multitude of soil microbes, not all of these can attack plants and some spoilage microbes are not common in soil, for example, lactic acid bacteria. Most spoilage losses are not due to microorganisms that cause plant diseases but rather to bacteria and molds that take advantage of mechanical and chilling damage to plant surfaces. Some microbes are found in only a few types of vegetables while others are widespread. *Erwinia carotovora* is the most common spoilage bacterium and has been detected in virtually every kind of vegetable. It can even grow at refrigeration temperatures.

Bacterial spoilage first causes softening of tissues as pectins are degraded and the whole vegetable may eventually degenerate into a slimy mass. Starches and sugars are metabolized next and unpleasant odors and flavors develop along with lactic acid and ethanol. Besides *E. carotovora*, several *Pseudomonas* spp. and lactic acid bacteria are important spoilage bacteria.

Molds belonging to several genera, including *Rhizopus, Alternaria*, and *Botrytis*, cause a number of vegetable rots described by their color, texture, or acidic products. The higher moisture content of vegetables as compared to grains allows different fungi to proliferate, but some species of *Aspergillus* attack onions.

Fresh Meat

In contrast to fruits and vegetables, meats are composed mainly of protein and fats rather than carbohydrates. Water content is 71–76% and therefore moisture is not an issue except for spoilage microbes on cured meats. Muscles of healthy animals do not contain any bacteria or fungi but as soon as animals are slaughtered, meat is exposed to contaminants and good sanitation practices are essential to produce high quality meats. The number of spoilage organisms on meat just after slaughter is a critical factor in determining shelf life. The surface of beef carcasses may contain anywhere from 10^1 to 10^7 cfu/cm^2, most of which are psychrotrophic bacteria. Chopping and grinding of meats can increase the microbial load as more surface area is exposed and more water and nutrients become available. A large variety of microbes are commonly found on fresh meat, but different microbes become dominant during spoilage of different meats depending on pH, composition and texture of processed meats, temperature and packaging atmosphere.

Pseudomonas spp. are the predominant spoilage bacteria in aerobically stored raw meat and poultry. Once the initial low levels of glucose are depleted by various microbes, *Pseudomonas* has an advantage because it can catabolize gluconates and amino acids more readily than other microbes. Break-down of these compounds results in production of malodorous sulfides, ammonia, and amines, including the biogenic amines putrescine and cadaverine. Dark, firm and dry meat with a relatively high pH of 6.0 spoils more rapidly because deamination of amino acids starts earlier. *Shewanella putrefaciens* does not grow on meat at pH<6.0 but can produce sulfides and ammonia even when glucose is still available. These sulfides not only smell bad but also cause color changes in meat, and therefore *Shewanella* has a high spoilage potential on fresh, high pH meats stored aerobically even when it is not a dominant microbe. *Brochothrix thermosphacta* is often a significant spoilage organism on fresh meat stored aerobically at refrigeration temperatures.

Enterobacteriaceae, particularly species of *Serra-tia, Enterobacter*, and *Hafnia*, are major causes of spoilage in vacuum-packed, high pH fresh meats. These organisms are facultative anaerobes that produce organic acids, hydrogen sulfide and greening of meats.

Lactic acid bacteria (LAB) grow on meat and poultry packaged under vacuum and modified atmospheres, producing organic acids from glucose by fermentation. This gives rise to aciduric off-odors which may be accompanied by gas and slime formation and greening of meat. However, LAB are weakly prote-olytic and so do not produce large amounts of amines or sulfides, and spoilage of meat by LAB is not as offensive.

Psychrophilic, anaerobic *Clostridium* spp. are associated with spoilage of vacuum-packaged meats. "Blown pack" meat spoilage is characterized by excessive gas formation with off odors due to formation of butyric acid, butanol and sulfurous compounds.

Yeasts and molds grow relatively slowly on fresh meat and do not compete well with bacteria. Therefore, they are a minor component of spoilage flora.

Processed Meat

Addition of sodium chloride, nitrites and/or nitrates, along with various other seasonings, emulsifiers and preservatives to ground or whole muscle meats changes the environment significantly and also the spoilage flora of processed meats. Dried and dry-fermented meats generally do not support microbial growth although process deviations may allow growth of some organisms. Spoilage organisms can grow on fresh and cooked cured meats, so they are best stored chilled, under a vacuum or modified atmosphere.

Pseudomonas spp. are not usually important causes of spoilage in processed meats because of their sensitivity to curing salts and heat pasteurization and their inability to grow well in meats packed with a vacuum or high carbon dioxide atmosphere. However, when packages have been opened and there has been insufficient curing, these bacteria may spoil refrigerated processed meats. Some cold- and salt-tolerant Enterobacteriaceae have been found to cause spoilage in some specific processed meats, such as ham or bacon.

Lactic acid bacteria (LAB) is the group of bacteria primarily associated with spoilage of processed meats. They produce sour off-flavors, gas, slime, and greening, and this spoilage may be more severe than in fresh meat because of the presence of added carbohydrates. Competitive ability of different LAB strains is related to pH and water activity of the meat, cooking and storage temperatures and oxygen and carbon dioxide levels.

Sporeformers (*Clostridium* and *Bacillus*) are usually not a spoilage problem in processed meats because of the presence of nitrite and other curing salts. However, faulty cooking/cooling procedures, including long cooling periods and temperature abuse, has allowed growth of these organisms in some cases. Spores of these organisms may be introduced with spices or other ingredients.

Yeasts cause some spoilage in processed meats but are generally only important when sulfite is used as a preservative or when meats have been irradiated or are stored aerobically in the cold. Slime may be produced along with vinegary or malty off-odors in some sausages.

Fish

Fish is a very perishable, high-protein food that typically contains a high level of free amino acids. Microbes metabolize these amino acids, producing ammonia, biogenic amines such as putrescine, hista-mine, and cadaverine, organic acids, ketones, and sulfur compounds. Degradation of lipids in fatty fish produces rancid odors. In addition, marine fish and some freshwater fish contain trimethylamine oxide that is degraded by several spoilage bacteria to trimethylamine (TMA), the compound responsible for fishy off odors. Iron is a limiting nutrient in fish, and this favors growth of bacteria such as pseudomonads that produce siderophores that bind iron.

Spoilage bacteria differ somewhat for freshwater and marine fish and for temperate and tropical water fish. Storage and processing conditions also affect microbial growth. *Pseudomonas* and *Shewanella* are the predominant species on chilled fresh fish under aerobic conditions. Packing under carbon dioxide and addition of low concentrations of sodium chloride favor growth of lactic acid bacteria and

Photobacterium phosphoreum. Heavily wet-salted fish support growth of yeasts while dried and salted fish are spoiled by molds. Addition of organic acids selects for lactic acid bacteria and yeasts. Pasteurization kills vegetative bacteria but spores of *Clostridium* and *Bacillus* survive and may grow, particularly in unsalted fish.

Fruits and Juices

Intact, healthy fruits have many microbes on their surfaces but can usually inhibit their growth until after harvest. Ripening weakens cell walls and decreases the amounts of antifungal chemicals in fruits, and physical damage during harvesting causes breaks in outer protective layers of fruits that spoilage organisms can exploit. Molds are tolerant of acidic conditions and low water activity and are involved in spoilage of citrus fruits, apples, pears, and other fruits. *Penicillium, Botrytis,* and *Rhizopus* are frequently isolated from spoiled fruits. Yeasts and some bacteria, including *Erwinia* and *Xanthomonas*, can also spoil some fruits and these may particularly be a problem for fresh cut packaged fruits. Fruits juices generally have relatively high levels of sugar and a low pH and this favors growth of yeasts, molds and some acid-tolerant bacteria. Spoilage may be manifested as surface pellicles or fibrous mats of molds, cloudiness, and off-flavors. Lack of oxygen in bottled and canned drinks limits mold growth. *Saccharomyces* and *Zygosaccharomyces* are resistant to thermal processing and are found in some spoiled juices. *Alicyclobacillus* spp., an aci-dophilic and thermophilic spore-forming bacteria, has emerged as an important spoilage microbe, causing a smoky taint and other off-flavors in pasteurized juices. *Propionibacterium cyclohexani-cum*, an acid-tolerant non-sporeforming bacterium also survives heating and grows in a variety of fruit juices. Lactic acid bacteria can spoil orange and tomato juices, and some pseudomonads and Entero-bacteriaceae also spoil juices. These bacteria are not as heat tolerant but may be post-pasteurization contaminants.

Spoilage of Foods by Moulds

Mould growth on foods, with its fuzzy or cottony appearance, sometimes coloured, is familiar to everyone, and usually food with a moldy or "mildewed" food is considered unfit to eat. While it is true that molds are involved in the spoilage of may kinds of foods, special molds are useful in the manufacture of certain foods or ingredients of foods. Thus, some kinds of cheese are moldripened, e.g., blue, Roquefort, Camembert Brie, Gammelost, etc., and molds are used in making Oriental foods, e.g., soy sauce, miso, sonti, and others discussed later. Molds have been grown as food or feed and are employed to produce products used in foods, such as amylase for breadmaking or citric acid used in soft drinks. Some molds do produce various toxic metabolites.

Moulds are microscopic, plant-like organisms, composed of long filaments called hyphae. Mould hyphae grow over the surface and inside nearly all substances of plant or animal origin. Because of their filamentous construction and consistent lack of chlorophyll they are considered by most biologists to be separate from the plant kingdom and members of the kingdom of fungi. They are related to the familiar mushrooms and toadstools, differing only in not having their filaments united into large fruiting structures.

Probably the best known microorganisms, molds are widely distributed in nature and grow under a variety of conditions in which air and moisture are present. They are also plants and a part of the fungi family. Nearly everyone has seen mold growth on damp clothing and old shoes. So, many may find it hard to believe that mold is a microorganism. However, the mold we see with the naked eye is actually a colony of millions of mold cells growing together. Molds vary in appearance. Some are fluffy and filament-like; others are moist and glossy; still others are slimy.

Unlike bacteria, molds are made up of more than one cell. Vegetative cells sustain the organism by taking in food substances for energy and the production of new cell material. Reproductive cells produce small "seed" cells called spores. Unlike bacterial spores, mold spores are the source of new mold organisms. Bacterial spores generally form only when environmental conditions are unfavourable. Molds produce a stem consisting of several cells. Together, these cells form a "fruiting body." The fruiting body produces the spores, which detach and are carried by air currents and deposited to start new mold colonies whenever conditions are favourable.

Mold spores are quite abundant in the air. So, any food allowed to stand in the open soon becomes contaminated with mold if adequate moisture is present. Some types of molds are also psychrophiles and can cause spoilage of refrigerated foods.

Molds are important to the food industry. Among their many contributions are the flavour and color they add to cheeses and the making of soy sauce. They also play a role in the making of such chemicals as citric and lactic acid and many enzymes. Molds can also cause problems in foods. Certain kinds can produce poisons called mycotoxins. Mycotoxins have only recently been discovered and little is known about what causes molds to produce them. Probably the best known use of molds is in the drug industry, where they help produce such antibiotics as penicillin.

Molds are plants of the kingdom Myceteae. They have no roots, stems, or leaves and are devoid of chlorophyll. They belong to the Eumycetes, or true fungi, and are subdivided further to subdivisions, classes, orders, families, and genera. The following criteria are used chiefly for differentiation and identification of molds:

— Hyphae septate or nonseptate

— Mycelium clear or dark (smoky)

— Mycelium coloured of colourless
— Whether sexual spores are produced and the type: oospores, zygospores, or ascospores
— Type of asecual spores: sporangiospores, conidia, or arthrospores (oidia)
— Charateristics of the spore head
  — Sporangia : size, color, shape, and location
  — Spore heads bearing conidia: single conidia, chains, budding conidia, or masses; shape and arrangement of sterigmata or phialides; gumming together of conidia
— Appearance of sporangiophores or conidiophores: simple or branched and if branched the type of branching; size and shape of columella at tipe of sporangiophore; whether conidiophores are single or in bundles.
— Microscopic appearance of the asexual spores, sepecially or conidia: shape, size, color; smooth or rough; one-, two-, or many-celled.

Presence of special structures (or spores): stolons, rhizoids, foot cells apophysis, chalamydospores, sclerotia, etc. Several molds of industrial importance are outlined by genus, with typical morphological structures presented in the accompanying figures.

Mucor

Mucor are involved in the spoilage of some foods and the manufacture of others. A widely distributed species is M. racemosus. M. rouxii is used in the “Amylo” process for the saccharification of starch, and mucors help ripen some cheese and are used in making certain Oriental foods.

Zygorrhynchus: These soil molds are similar to Mucor except that the zygospore suspensors are markedly unequal in size.

Rhizopus: Rhizopus stolonifer, the so-called bread mold, is very common and is involved in the spoilage of may foods: berries, fruits, vegetables, bread, etc. Rhizopus is closely related to Mucor and inhabits the same ecological niches. R. nigricans is one of the most common members of the Mucorales and has a worldwide distribution, although most commonly occurring in warmer areas. It has frequently been reported from dry habitats. The spores are dispersed in hot, dry weather. It is frequently isolated from both forest and cultivated soils. It occurs in children’s sandboxes. Typical microhabitats include fresh or decaying litter such as pine needles and leaves. Other known substrates are sweet potato, cold-stored straw- berries, stewed fruits, and the nests, feathers and droppings of wild birds.

Rhizopus is often included in the list of molds which are thought to have clinical relevancy for skin testing mold-sensitive patients. The risk for occupational exposure is most likely to occur among food handlers during the storage, transfer and marketing of strawberries, peaches, cherries, corn and peanuts.

Absidia: Similar to Rhizopus, except that sporangia are small and pearshaped.

Thamnidium: elegans is found on meat in chilling storage, causing "whiskers" on the meat.

Aspergillus: The aspergilli are very widespread. Many are involved in the spoilage of foods, and some are useful in the preparation of certain foods. Raper and Fennell list eighteen groups of aspergilli and recognise 132 species, but only a few will be mentioned here.

The A. glaucus group, with A. repens as an important species, is often involved in food spoilage. The molds grow well in high concentrations of sugar and salt and hence in many foods of low moisture content. Conidia of this group are some shade of green, and ascospores are in asci within yellow to reddish perithecia. Some authors include these molds in the Ascomycetes and the genus Eurotium, a name reserved for members having a perfect stage.

The A. niger group, with A. niger as a leading species, is widespread and may be important in foods. The spore-bearing heads are large, tightly packed, and globular and may be black, borwnish-black, or purple-brown. The conidia are rough, with bands of pigment. Many strains have scelerotia, coloured from buff to gray to blackish. Selected strains are used for the commercial production of citric and gluconic acids and in a variety of enzyme preparations. The A. flavus-oryzae group includes molds important in the making of some Oriental foods and the production of enzymes, but molds in this group often are involved in the spoilage of foods. Conidia give various yellow to green shades to the spore heads, and dark sclerotia may be formed.

Penicillium: This is another genus that is widespread in occurrence and important in foods. The genus is divided into groups and subgroups, and there are numberous species. The genus is divided into large groups on the basis of the branching of the spore-bearing heads, or penicilli. These heads, or verticillata, are a whorl or cluster of three or more elements: sterigmata, metulae (subbranches), and branches. P. expansum, the blue-green-spored mold, causes soft rots of fruits. Other important species are P. digitatum, with olive-or yellowish-green conidia, causing a soft rot of citrus fruits; P. italicum, called the "blue contact mold," with blue-green conidia, also rotting citrus fruit; P. camemberti, with grayish conidia, useful in the ripening of Camembert cheese, and P. roqueforti, with bluish-green conidia, aiding in the ripening of blue cheeses, e.g., Roquefort. A few species form asci with ascospores in cleistothecia, and a few exhibit sclerotia and therefore have caused trouble in canned acid foods. 1-monovertcillata symmetrica, 2-monoverticillata symmetrica, biverticllata symmetrica, 4-polyverticillata symmetrica.

Trichothecium: The common species, T. roseum, is a pink mold which grows on wood, paper, fruits such as apples and peaches, and vegetables such as cucumbers and cantaloupes. This mold is easily recognised by the clusters of two-celled conidia at the ends of short, erect conidiophores. Conidia have a nipplelike projection at the point of attachment, and the smaller of the two cells of each conidium is at this end.

Geotrichum: This genus is included with the yeastlike fungi by some writers and with the molds by others. Species may be white, yellowish, orange, or red, with the growth appearing first as a firm, feltlike mass that later becomes soft and creamy. Geotrichum candidum, often called the "dairy mold." Gives white to creamcolored growth. The hyphae are septate and in common species are arthrospores (oida), which may appear rectangular if from submerged hyphae and oval if from aerial hyphae.

Neurospora: This genus has been described under various names because of the confusion concerning its classification, but most mycologists believe that it should be classed among the perfect molds (producing sexual spores) and call the genues Neurospora. Neurospora (Monilia) sitophila, the most important species in foods, sometimes is termed the "red bread mold" because its pink, loose-textured growth often occures on bread. It also grows on sugarcane bagasse and on various foods. The perfect, or ascosporogenous stage seldom is seen.

Sportotrichum: Among the saprophytic species is S. carnis, found growing on chilled meats, where it causes "white spot."

Botrytis: One species important in foods is B. Cinerea. It causes a disease of grapes but may grow saprophytically on many foods. B. cinerea has a worldwide distribution but occurs mainly in humid, temperate and subtropical regions. B. cinerea is found regularly in the soil, though its proportion of the total fungus population is not high. It can be parasitic on a wide range of plants, causing blight or rot of leaves, flowers and fruits. It is called the grey mold, as it covers the decayed tissues with conidiophores, e.g. grey mold of cabbage or lettuce, tomato. It is especially seen in connection with soft fruits, e.g. strawberries and grapes.

Cephalosporium: C. acremonium is a common species.

Trichoderma: T. viride is a common species. The mature mold plant is bright green because the balls of green conidia are glued together, and tufts of white hyphae (sterile) stick up well above the conidiophores.

Scopulariopsis: S. brevicaulis is a common species. This genus may be confused with Penicillium, for both have brushlike penicilli and chains of spores cut off from the sterigmata, but the conidia of Scopulariopsis are never green. Conidiophores may be branched or unbranched in Scopulariopsis, and simple or complex, and the branching usually is irregular. The spore-bearing heads may very from complex, branching systems with penicilli to single sterigmata arising from short branches of aerial hyphae. The spores are distinctive in microscopic appearance and are not green out commonly yellowish-brown; they are lemonshaped, thickwalled, spiny and pointed at one end, with a thick ring at the opposite end. Colonies are brownish and cottony.

Pullularia: Ovate, hyaline conidia (blastospors or buds from preexisting cells) borne as lateral buds on all parts of the mycelium. Colonies are pale and slimy and yeastlike when young, becoming mycelial and dark and leathery in age. P. pullalans is a common species.

Cladosporium: C. herbarum is a leading species. These dark molds cause "black spot" on a number of foods, on cellar walls, etc. Colonies of C. herbarum are restricted in growth and are thick, velvety, and olive—to gray-green; the reverse side of the plant is a striking opalescent blueblack to greenish-black. Cladosporium is the most frequently encountered mold in the air. The dry conidia are carried easily though the air and transported even over oceans. The conidia of Cladosprium can be detected in extremely large numbers. The indoor counts to a large extent reflect the outdoor concentration. Depending on climatic conditions the conidia may begin to appear in the atmosphere in the spring and rise to a peak in either late summer or early fall. Cladosporium is one of the most common colonisers of dying and dead plants and also occurs in various soil types, and on food. This mold is frequently found in uncleaned refrigerators, foodstuffs, on moist window frames, in houses with poor ventilation, with straw roofs, and situated in low, damp areas. It has been isolated from fuel tanks, face creams, paints and textiles.

Helminthosporium: Species of this genus are for the most part plant pathogens but may grow saprophytically on vegetable materials.

Alternaria: Molds of this genus are common causes of the spoilage of foods A. citri (rotting citrus fruits), A. tenuis, and A. brassicae are common species. The mass of mycelium usually is dirty gray-green, but hyphae often look nearly colourless under the microscope. The brown, many-celled conidia are in a chain on the conidiophore.

Stemphylium: This, too, is a common genus. The conidia are dark and multicellular but have fewer cross walls than those of alternaria and are rounded at both ends. S. botryosum is a cosmopolitan species, very common in temperate and subtropical regions. Soil isolates have been reported from forests, grasslands, wheat plantings, beet and citrus cultivation and coffee plan- rations. It has also been isolated from polluted fresh water, leaf litter of trees, bark and leaves of citrus. As a typical seedborne fungus, it is seen on tomato, wheat and barley. The conidia are violently discharged after a lowering of the relative humidity in the presence of light.

Fusarium: Molds of this genus often grow on foods. The species are very difficult to identify, and the appearance of growth is variable. F. moniliforme is a universal dominant found in aerobiologic surveys worldwide. it is widely distributed on numerous grasses and other plants and is a common soil fungus. It often causes plant diseases and is a major parasite of rice, sugar cane, and sorghum and is especially common on maize grains. It occurs regularly on banana roots and other fruits and vegetables, e.g. tomato and watermelon. It sporulates in warm, wet weather. During winter or in dry periods, the fungus survives in the soil and on plant debris.

Endomyces: Yeastlike fungi, forming mycelium and arthrospores. Some species rot fruits.

Monascus: Colonies of M. purpureus are thin and spreading and reddish or purple in colour. Found on dairy products and on Chinese red rice (ang-khak).

Sclerotinia: Some species causes rots of vegetables and fruits, where they are present in the conidial stage. The lemon shaped conidia are in chains, with a "plug" separating conidia. In the preceding discussion of molds important in foods, genera that only occasionally grow on specific foods have been omitted. They will be mentioned with those foods.

Microbial Spoilage of Cereal Grains

Cereal grains and meals and flours made from them should not be subject to microbial spoilage if they are prepared and stored properly because their moisture content is too low to support even the growth of molds. If, however, these products become moistened above the minimum for microbial growth, with follow. A little moistening will permit only growth of molds, but more moisture will allow the growth of yeasts and bacteria.

Although the microbial load of cereal grains, meals, and flours may not constitute a spoilage problem by itself, the numbers and types of microorganisms in such products are of concern since these products are used in the formulation of many other foods. The microbial contribution of cereal grains and flours to convenience foods is an important consideration from a public health aspect and as a source of possible spoilage agents.

Since cereal grains and meals ordinarily are not processed to reduce their natural flora of microorganisms greatly, they are likely to contain molds, yeasts, and bacteria, which are ready to grow if enough moisture is added. In addition to starch, which is unavailable to many organisms, these grains contain some sugar and available nitrogen compounds, minerals, and accessory growth substances; and the analyses will release more sugar and proteinases will yield more available nitrogenous foods if the grains are moistened. A little added moisture will result in growth of molds at the surface, where air is available.

A wet mash of the grains or a mash of the meals will undergo an acid fermentation, chiefly by the lactic acid and coliform bacteria normally present on plant surfaces. This may be followed by an alcoholic fermentation by yeasts as soon as the acidity has increased enough to favour them. Finally, molds (and perhaps film yeasts) will grow on the top surface, although acetic acid bacteria, if present, may oxidise the alcohol to acetic acid and inhibit the molds.

The major factors involved in the spoilage of stored grain by molds include microbial content, moisture levels above 12 o 13 percent, physical damage, and temperature. Numerous different molds can be involved, but the most common are species of Aspergillus, Penicillium, Mucor, Rhizopus, and Fusarium. As previously mentioned, many of these molds can produce mycotoxins. Grains spoiled by molds represent a potential animal or human health hazard as well as a great economic loss.

Role of Flours

Dry cleaning and washing grains and the milling and shifting of flour reduce the content of microorganisms, but the important kinds still are represented in wholegrain flours, e.g., whole wheat or buckwheat, and the spoilage would be similar to that described for cereal grains and meals.

White wheat flour, however, usually is bleached by an oxidising agent, such as an oxide of nitrogen, chlorine, nitrosyl chloride, or benzoyl peroxide, and this process serves to reduce microbial numbers and kinds. A moisture content of flour of less than 13 percent has been reported to prevent the growth of all microorganisms. Other workers claim that 15 percent permits good mold growth, and over 17 percent the growth of both molds and bacteria. Therefore, slight moistening of white flour brings about spoilage by molds. Because of the variations in microbial content of different lots of flour, the type of spoilage in a flour paste is difficult to predict.

If acid-forming bacteria are present, an acid there and then acetic acid by Acetobacter species. This succession of changes would be more likely in freshly milled flour than in flour that had been stored for a long period with a consequent reduction in kinds and numbers of microorganisms. In the absence of lactics and coliforms, micrococci have been found to acidify the paste, and in their absence species of Bacillus may grow, producing lactic acid, gas, alcohol, acetoin, and small amounts of esters and other aromatic compounds. It is characteristic of most flour pastes to develop an odor of acetic acid and esters.

Dough and Bread

Food spoilage,a very mouldy loaf of bread calcium propionate will not protect it forever. The fermentations taking place in the doughs for various kinds of bread will be discussed, where it will be noted hat some changes caused by microorganisms are desirable and even necessary in making certain kinds of bread. The acid fermentation by lactics and coliform bacteria that is normal in flour pastes or doughs may be too extensive if too much time is permitted, with the result that the dough and bread made from it may be too "sour."

Excessive growth of proteolytic bacteria during this period may destroy some of the gasholding capacity so essential during the rising of the dough and produce a sticky dough. Sticky doughs, however, are usually the result of overmixing or gluten breakdown by reducing agents, e.g., glutathione. There also is the possibility of the production by microorganisms of undesirable flavours other than the sourness.

Mold Spores

Molds are the most common and hence the most important cause of the spoilage of bread and most bakery products. The temperatures attained in the baking procedure usually are high enough to kill all mold spores in and on the loaf, so that molds must reach the outer surface or penetrate after baking. They can come form the air during cooling or thereafter, from handling, or from wrappers and usually initiate growth in the crease of the loaf and between the slices of sliced bread.

Chief molds involved in the spoilage of bread are the so-called bread mold, Rhizopus stolonifer, with its white cottony mycelium and back dots of sporangia; the greenspored Penicillium expansum or P. stoloniferum; Aspergillus niger with its greenish- or purplish-brown to black conidial heads and yellow pigment diffusing into the bread; and Monilia (Neurospora) sitophila, whose pink conidia give a pink or reddish color to its growth. Species of Mucor or Geotrichum or any of a large number of species of other genera of molds may develop. Mold spoilage is favored by

— heavy contamination after baking, due, for example, to air heavily laden with mold spores, a long cooling time, considerable air circulation, or a contaminated slicing machine,
— slicing, in that more air is introduced into the loaf,
— wrapping, especially if the bread is warm when wrapped, and
— storage in a warm, humid place.

There is little growth of commercial importance on bread crust in a relative humidity below 90 percent. Bread with 6 percent of milk solids retains moisture somewhat better than does milkfree bread, and hence there is less moisture between load and wrapper and hence less molding, but there is not enough of an effect to be of much practical importance. Molding often begins within a loaf of sliced bread, where more moisture is available than at the surface, especially in the crease.

Prevent moldiness of bread

Prevention of contamination of bread with mold spores insofar as practicable. The air about the bread is kept low in spores by removal of possible breeding places for molds, such as returned bred, or walls and equipment. Spore-laden flour dust from other parts of the bakery has been incriminated in causing increased moldiness of bread. Filtration and washing of air to the room and irradiation of the room and more especially the air by means of ultraviolet rays cut down contamination. Prompt and adequate cooling of the loaves before wrapping to reduce condensation of moisture beneath the wrapper. Ultraviolet irradiation of the surface of the loaf and of slicing knives.

Destruction of molds on the surface by electronic heating. Keeping the bread cool to slow mold growth or freezing and storage in the frozen condition to prevent growth entirely. Icorporation in the bread dough of some mycostatic chemical. Most commonly employed now is sodium or calcium propionate at the rate of 0.1 and 0.3 percent of the weight of the flour, a treatment that also is effective against rope. Sorbic acid, up to 0.1 percent, and sodium diacetate, up to 0.32 percent, are also used. An older remedy was the addition of vinegar or acetate to the dough or treatment of the exterior of the loaf with vinegar.

Rope formation

Ropiness of bread is fairly common in home-baked bread, especially during hot weather, but it is rare in commercially baked bread because of the preventive measures now employed. Ropiness is caused by a mucoid variant of Bacillus subtilis or B. licheniformis, formerly called B. mesentericus, B. panis, and other speices names. The spores of these species can withstand the temperature of the bread during baking, which does not exceed 100 C, and can germinate and grow in the loaf if conditions are favorable.

The ropy condition apparently is the result of capsulation of the bacillus, together with hydrolysis of the flour proteins (gluten) by proteinases of the organism and of starch by analyse to give sugars that encourage rope formation. The area of ropiness is yellow to brown in color and is soft and sticky to the touch. In one stage the slimy material can be drawn out into long threads when the bread is broken and pulled apart. The unpleasant odor is difficult to characterise, although it has been described as that of decomposed or overripe melons. First the odor is evident, then discoloration, and finally softening of the crumb, with stickiness and stringiness.

Red bread

Red, or "bloody," bread is striking in appearance but rare in occurrence. The red color results from the growth of pigmented bacteria, usually Serratia marcescens, an organism that often is brilliantly red on starchy foods. In ancient times the mysterious appearance of apparent drops of blood was considered miraculous. Necessary for the phenomenon is the accidental contamination of the bread with the red organisms and unusually moist conditions to favor their growth. Molds, such as Monilia (Neeurospora) sitophila, previously mentioned, may impart a pink to red color to bread. A red color in the crumb of dark bread has been caused by Geotrichum aurantiacum.

Chalky bread

Chalky bread, also uncommon, is so named because of white, chalklike spots. The defect has been blamed on the growth of yeast like fungi, Endomycopsis fibuligera nad Trichosporon variable.

Bakery Products

Molds are the chief cause of the microbial spoilage of cakes and other bakers' products, since the normal baking process destroys much of the original microflora. Methods of prevention are similar to those previously listed for bread. The microbiology of these products becomes more complicated when they are topped with frostings or fruits or filled with custard, imitation creams, or sauces.

Usually the toppings or fillings are more prone to microbial spoilage than is the actual baked portion. It is not unusual for the fillings of many pasteries to support the growth of microorganisms. Frostings, because of their high sugar content, are quite stable but may be spoiled by molds or yeast upon storage. The deterioration of breads, cakes, pies, and other bakery products that is referred to as staling is due mostly to physical changes during holding and not to microorganisms.

Pasta here is used to describe egg-based pasts usually containing flour, water, and eggs. Pasta is delivered and stored dry; therefore, there are seldom reports of spoilage in these products. Macaroni usually contains only flour, water, and other nutrients. The swelling of moist macaroni has been reportedly caused by gas production by bacteria resembling Enterobacter cloacae. During the drying of macaroni on paper a mold of the genus Monilia has been found responsible for purple streaks at contact points with the paper. The appearance of these defects is uncommon, however, despite the long, slow drying process of macaroni.

Tapioca, prepared from the root starch of cassava, will spoil if moistened. Spoilage by an orange-pigmented, starchhydrolysing bacterium has been described.

According to Hobbs and Green, there are three basic breakfast cereal manufacturing processes: These products are flaked, puffed, or extruded. In the initial steps of the manufacturing of these products there are high levels of moisture and, therefore, the possibility of microbial growth. However, in the finished state they usually represent products with low total numbers.

A large number of products available at retail are essentially prepared, refrigerated, or frozen bakery product doughs. These convenience items may contain a yeast or a lactic acid bacteria inoculum. The level if contamination and the subsequent microbial quality would be a direct result of the quality of ingredients used and the sanitary practices employed during their manufacture. Microbial numbers may increase during refrigerated storage until the product receives its final "heat treatment" at the consumer level.

The microbial flora of fresh, unsterile, dough products held at refrigeration temperatures was compared with the microbial flora of the same products that had spoiled spontaneously. Various methods based on selective media were used to determine molds, yeasts, and bacteria present. Except for two special cases in which a yeast and Penicillium roqueforti induced spoilage, all of the samples deteriorated because of bacterial growth. A total of 1,132 bacterial isolates was subjected to further classification.

In the spoiled products, 92% of the isolates belonged to the Lactobacillaceae. More than one-half of these (53%) belonged to the genus Lactobacillus, and an additional 36% were in the genus Leuconostoc. In the genus Leuconostoc almost all of the strains (94%) were L. mesenteroides. The third most common genus present was Streptococcus, represented by

3% of the total isolates. A preliminary taxonomic study of the microflora of refrigerated dough products revealed none of the isolates to be indicators of fecal contamination and none to be forms known to produce toxins.

Spoilage of Fruits and Vegetables

In some countries, chemical disinfectant agents are used to decontaminate the surface of fruits and vegetables, in addition to washing with water. It is not clearly known to which extent these agents are effective, what their optimum conditions of use are, and whether they have adverse toxicological effects.

Fruits and vegetables eaten raw, have long been known to serve as vehicles for transmission of infectious microorganisms in developing countries. The number of confirmed cases of illness associated with raw fruits and vegetables in industrialised countries, on the other hand, has been relatively low compared to the number of recorded cases due to foods of animal origin. The impact of agronomic practices, handling, processing, distribution and marketing on risks associated with microbiological hazards of raw fruits and vegetables is not well understood.

Disinfectants are effective in reducing the bacterial load but their efficacy depends on the types of fruits and vegetables and characteristics of their surface, the method and the procedure used for disinfection, and on the type of pathogen. Listeria monocytogenes is generally more resistant to disinfectants than Salmonella, pathogenic Escherichia coli and Shigella, but little is known about the efficacy of disinfectants in killing parasites and viruses on fruits and vegetables.

Vigorously washing fruits and vegetables with safe water reduces the number of microorganisms by 10-100-fold and is often as effective as treatment with 200 ppm chlorine. Treatments with chlorine dioxide, trisodium phosphate, organic acids or ozone have potential for removing pathogenic microorganisms from raw fruits and vegetables. When applicable, food irradiation is one of the most powerful methods of decontamination.

Prevention of contamination at all points of the food chain from primary production to the final consumer is preferred over the application of disinfectant after contamination occurs. Prevention can be achieved through application of the principles of food hygiene and the Hazard Analysis and Critical Control Point system.

Research is needed to better understand not only the mechanisms through which pathogens can contaminate raw fruits and vegetables but also the procedures for killing or removing pathogens once they are present, either on the surface or in internal tissues, and the analytical methods for their detection. A quantitative microbiological risk assessment should be undertaken of human infections and intoxications that can be linked to the consumption of contaminated raw fruits, vegetables and plant materials.

The development of a highly efficient, international epidemiological surveillance system for better understanding the role of raw fruits and vegetables as vehicles for infectious diseases is critical. Information generated by such a system would be valuable in establishing more meaningful practices and guidelines for preventing contamination and for decontaminating raw fruits and vegetables.

Continual training of fruit and vegetable growers and handlers at all levels should be carried out in order to control microbiological hazards that may be influenced by current and changing practices in aquaculture, agronomy, processing, marketing and preparation.

The use of water and fertilisers free of pathogenic microorganisms at all stages of growing and handling and, just as importantly, an appropriate level of understanding of the importance of personal hygiene by everyone involved is essential if the risk of illness is to be minimized. Regardless of the stringency of hygienic measures at production and processing stages of the food chain, consumers should be educated to realise the importance of adequate washing of fruits and vegetables.

Raw fruits and vegetables have been known to serve as vehicles of human disease for at least a century. In 1899, Morse linked typhoid infection to eating celery. Warry attributed an outbreak of typhoid fever to eating watercress grown in soil fertilised with sewage and Pixley recorded two cases of typhoid from eating uncooked rhubarb which was grown in soil known to have been fertilised with typhoid excreta. In 1912, Creel demonstrated that lettuce and radishes grown in soil containing Bacillus typhosa harboured the organism on their surfaces for up to 31 days.

Melick recovered typhoid bacilli from mature lettuce and radish harvested from soil that had been inoculated at the time seeds were planted. Some parasitic helminths require encystment on plants to complete their life cycle. Thus, the recognition of raw fruits and vegetables as potential vehicles for transmission of pathogenic microorganisms known to cause human disease is not new.

Nevertheless, documented outbreaks of foodborne illness associated with fruits and vegetables in industrialised countries are relatively rare. For instance, in 1996 only six of about 200 reported foodborne disease outbreaks in the United Kingdom were associated with consumption of fruits and vegetables. In recent years, however, the frequency of outbreaks epidemiologically associated with raw fruits and vegetables is documented to have increased in some industrialised countries as a result of change in dietary habits and increased import of food.

In developing countries, foodborne illnesses caused by contaminated fruits and vegetables are frequent and in some areas they cause a large proportion of illness. However, due to lack of foodborne disease investigation and surveillance in most of these countries, most outbreaks go undetected and the scientific literature reports only on very few outbreaks. In 1995-1996, only 2% of foodborne disease outbreaks in Latin America were related to fruits and vegetables. Raw and minimally-processed fruits and vegetables are an essential part of people's diet all around the world. Where land is available, families grow fruits and vegetables for their own use.

Alternatively, produce is purchased from local farmers or retail outlets for further preparation by street vendors, by families at home or as part of meals eaten in restaurants and other food service facilities. While advances in agronomic practices, processing, preservation, distribution and marketing have enabled the raw fruit and vegetable industry to supply highquality produce to many consumers all year round, some of these same practices have also expanded the geographical distribution and incidence of human illness associated with an increasing number of pathogenic bacterial, viral and parasitic microorganisms.

Changes that may contribute to the increase in diseases associated with the consumption of raw fruits and vegetables in industrialised countries and foods in general have been described.

Factors include globalisation of the food supply, inadvertent introduction of pathogens into new geographical areas,e.g. outbreaks of shigellosis in Norway, Sweden and the United Kingdom in 1994 due to contaminated lettuce imported from southern Europe and cyclosporiasis in the United States which was linked to consumption of contaminated raspberries imported from Guatemala and the development of new virulence factors by microorganisms, decreases in immunity among certain segments of the population, and changes in eating habits. In developing countries, continued use of untreated wastewater and manure as fertilisers for the production of fruits and vegetables is a major contributing factor to contamination that causes numerous foodborne disease outbreaks.

While every effort should be made to prevent contamination of fruits and vegetables during production, transport, processing and handling, much improvement is still needed in some parts of the world if hygienic production of fruits and vegetables is to be ensured.

Recognising the problem

Fruits and vegetables can become contaminated with microorganisms capable of causing human diseases while still on the plant in fields or orchards, or during harvesting, transport, processing, distribution and marketing, or in the home. Bacteria such as Clostridium botulinum, Bacillus cereus and Listeria monocytogenes, all capable of causing illness, are normal inhabitants of many soils, whereas Salmonella, Shigella, Escherichia coli and Campylobacter reside in the intestinal tracts of animals, including humans, and are more likely to contaminate raw fruits and vegetables through contact with faeces, sewage, untreated irrigation water or surface water.

Contamination may also occur during post-harvest handling, including at points of preparation by street vendors, in foodservice establishments and in the home. Contamination with viruses or parasites can result from contact with faeces, sewage and irrigation water.

Numerous surveys have been carried out in many countries to determine the presence of pathogenic microorganisms on raw fruits and vegetables. In many instances, bacterial pathogens have not been detected. In other investigations, high percentages of samples have been found to contain bacteria capable of causing human disease. This list is not comprehensive but it reflects the scope of potential problems that may link raw fruits and vegetables to human disease.

Indeed, numerous outbreaks of illness caused by bacteria, viruses and parasites have been linked epidemiologically to consumption of raw fruits and vegetables. Increased recognition of raw fruits and vegetables as suspected vehicles of human illness in industrialised countries may probably result in increased numbers of associated or confirmed cases in the future.

Improved diagnostic systems and surveillance programmes will enhance the prospect of identifying raw fruits and vegetables as sources of foodborne illness, thus possibly resulting in an increase in recorded numbers of outbreaks and cases of disease. Surveys of raw fruits and vegetables for the presence of parasites or viruses are few, largely because of the lack of sensitive methods for their detection in plant materials.

Other factors, however, may contribute to real increases in diseases associated with fruits and vegetables. These include use of wastewater, increased application of improperly composted manures to soils in which fruits and vegetables are grown, changes in packaging technology such as the use of modified or controlled atmosphere and vacuum packaging, extended time between harvesting and consumption, and changing food consumption patterns.

Increased global trade in raw fruits and vegetables, as well as increased international travel in general, could also increase the risk of produce-associated diseases. Finally, the susceptibility of the public to foodborne diseases, at least in more developed countries, is changing due to increased numbers of people who are elderly, immunocompromised or have chronic diseases. This change in social demographics is likely to lead to increased risk of illness associated with the consumption of raw produce that otherwise may contain levels of pathogens innocuous to healthy individuals.

Prevention of contamination of fruits and vegetables with pathogenic microorganisms should be the goal of everyone involved in both the pre-harvest and post-harvest phases of delivering produce to the consumer. This is a very difficult task, since some pathogens are normally present in the soil and may therefore be present on the surface of fruits and vegetables when they are harvested.

The use of properly composted manure and properly treated irrigation and spray waters, as well as pathogen-free water for washing, and for ice, will minimise the risk of contamination of fruits and vegetables with microbial pathogens. Good hygienic practice during production and transport, including sanitising of harvesting equipment and transport vehicles, as well as the application of good hygienic practice during processing and preparation are critical. Elimination of animals and insects from processing, storage, marketing and foodservice facilities should be a goal of anyone who handles raw produce. The highest level of hygiene must be practised by all handlers (including consumers) of fruits and vegetables, from the field to the table, if any degree of success is to be achieved in minimising the risk of contamination.

The microorganisms normally present on the surface of raw fruits and vegetables may consist of chance contaminants from the soil or dust, or bacteria or fungi that have grown and colonised by utilising nutrients exuded from plant tissues. Among the groups of bacteria commonly found on plant vegetation are those that test positive for coliforms or faecal coliforms,e.g. Klebsiella and Enterobacter. Thus, the presence of coliforms or faecal coliforms on raw fruits and vegetables does not necessarily provide an index of faecal contamination.

Microorganisms capable of causing human disease can, however, be found on raw produce, and should be viewed as a threat to public health. What is not well known is where contamination occurs along the farm-to-fork pathway. Means to address this issue must address the entire system from production to consumption.

The presence of pathogenic and other microorganisms on fruits and vegetables is dictated not so much by surface topography as by exposure to environmental factors that lead to contamination. Although rough, highly-textured surfaces with deep crevices would be more likely to harbour soil, with the possible consequence of increased numbers of microorganisms, than would smooth-surfaced fruits and vegetables, differences in microbial profiles result largely from unrelated factors such as resident microflora in the soil, application of nonresident microflora via animal manures, sewage or irrigation water, rainfall and atmospheric humidity.

The presence of a pathogen on produce is of less consequence if the rind, skin or peel is to be removed before consumption. Bananas, melons, mangoes, durians, pineapples and papayas, for example, fall in this category. However, the process of removing the rind, skin or peel from these fruits, perhaps with the exception of bananas, may result in contamination of the edible portion, thereby creating a risk to the consumer. Microorganisms that have become trapped on the inner leaves of certain vegetables can be particularly difficult to remove by routine cleansing practices.

Most Concern Pathogens

Salmonella and untreated sewage

The antigenic scheme for classifying salmonellae recognises more than 2300 serovars and, while all can be considered human pathogens, only about 200 are associated with human illness. Animal husbandry practices used in the poultry, meat and fish industries, and the recycling of offal and inedible raw materials into animal feeds, has favoured the continued prominence of Salmonella in the global food chain. There are reports of human salmonellosis linked to cantaloupe and sprouts produced from alfalfa seeds imported to the United States.

Hygienic conditions during the production, harvesting, transport and distribution of raw fruits and vegetables from some countries may not always meet minimum hygienic requirements, thus facilitating contamination on arrival in another country. Application of night soil, untreated sewage sludge or effluents, or irrigation water containing untreated sewage to fields and gardens can result in contamination of fruits and vegetables with Salmonella and other pathogens. Washing fruits and vegetables with contaminated water and handling of produce by infected workers, vendors and consumers in the marketplace helps the spread of pathogenic microorganisms, including Salmonella.

Salmonellae have been isolated from many types of raw fruits and vegetables. Outbreaks of salmonellosis have been linked to a diversity of fruits and vegetables, including tomatoes, bean sprouts, melons, unpasteurised orange juice and apple juice. The pathogen can grow on the surface of alfalfa sprouts, tomatoes and perhaps on other mature raw fruits and vegetables, making it imperative to use hygienic practices when handling them.

Shigella

Bacillary dysentery or shigellosis is caused by Shigella, of which there are four species: S. dysenteriae, S. flexneri, S. boydii and S. sonnei. Most cases of shigellosis result from the ingestion of food or water contaminated with human faeces. Like salmonellae and other pathogens present in faeces, Shigella can contaminate raw fruits and vegetables by several routes, including insects and the hands of persons who handle the produce, although shigellosis is more often transmitted from person to person.

Several large outbreaks of shigellosis have been attributed to the consumption of contaminated raw fruits and vegetables. Lettuce, scallions, vegetable salad, potato salad containing spring onions, salad vegetables and watermelon have been implicated as vehicles of shigellosis. Sliced raw papaya, jicama and watermelon support the growth of Shigella species. S. sonnei survived on refrigerated, shredded lettuce for three days without decreasing in number, and increased when held at 22°C.

Escherichia Coli

Escherichia coli is common in the normal microflora of the intestinal tracts of humans and other warm-blooded animals. Strains that cause diarrhoeal illness are categorised into groups on the basis of virulence properties, mechanisms of pathogenicity, clinical syndromes and antigenic characteristics. The major groups are designated as enterotoxigenic, enterohaemorrhagic, enteropathogenic, enteroinvasive, diffuseadhering and enteroaggregative.

Fruits and vegetables can become contaminated with one or more of these groups while in the field or during post-harvest handling. Sources and mechanisms of contamination are similar to those described for Salmonella and Shigella. Enterotoxigenic E. coli is a cause of traveller's diarrhoea, an illness sometimes experienced when individuals visit countries with food and water hygiene standards different from their own.

Contaminated raw vegetables are thought to be a common cause of traveller's diarrhoea. Illness has been associated with consumption of salads and carrots. Enterohaemorrhagic E. coli O157:H7 has more recently been recognised as a foodborne pathogen. Since cattle appear to be a natural reservoir for the pathogen, most outbreaks of illness have been associated with the consumption of contaminated, undercooked beef and dairy products. However, outbreaks have also been linked to lettuce, apple cider, radish sprouts and alfalfa sprouts. Enterohemorrhagic E. coli can grow on cantaloupe and watermelon cubes, shredded lettuce and sliced cucumbers, and in apple cider.

Contamination of raw fruits and vegetables with enterohaemorrhagic E. coli O157:H7 may occur when cattle, and perhaps other ruminants such as deer, inadvertently enter fields, or when improperly composted cow manure has been applied as a fertiliser. The potential for contamination may be enhanced when fruits or vegetables have fallen from the plant to the ground and are then picked and placed into the handling and processing chain. Also, because contaminated manure may become airborne dust particles, it is possible that fruits on trees and vines may become contaminated. Workers on farms and in packing houses may also be a source of E. coli O157:H7. These mechanisms of contamination are somewhat speculative at present and must be thoroughly investigated before appropriate interventions can be introduced to reduce the risk.

Campylobacter

Campylobacter jejuni is a leading cause of bacterial enteritis in many countries. Reservoirs of this pathogen include several wild animals as well as poultry, cows, pigs and domestic pets. While consumption of food of animal origin, particularly poultry, is largely responsible for infection, Campylobacter enteritis has also been associated with the consumption of raw fruits and vegetables. Although Campylobacter does not grow at temperatures below 30 °C and is sensitive to acid pH, it can survive on cut fruits for sufficient time to be a risk to the consumer.

Yersinia enterocolitica

Yersinia enterocolitica can be found in a variety of terrestrial and freshwater ecosystems, including soil, vegetation and water in lakes, rivers, wells and streams, but most isolates from these sources lack virulence for humans. Pigs, however, frequently carry serotypes capable of causing human disease. The ability of Y. enterocolitica to grow at refrigeration temperature and its documented presence on raw produce raises concern about the potential of salad vegetables as causative vehicles of yersiniosis in humans.

Seven per cent of carrot samples obtained from eating establishments in France were reported to contain serotypes of Yersinia that may be pathogenic to humans. In another study of raw vegetables analysed contained nonpathogenic strains of Yersinia. Incidence was higher on root and leafy vegetables than on tomatoes or cucumbers. Certainly, application of improperly composted pig manure to vegetable fields should be avoided to reduce the possibility of pathogenic strains being present on produce when it reaches the consumer.

Listeria monocytogenes

Listeria monocytogenes is present in the intestinal tract of many animals, including humans, so it is not surprising that the organism can also be found in the faeces of these animals, on the land they occupy, in sewage, in soils to which raw sewage is applied and on plants which grow in these soils. The organism also exists in nature as a saprophyte, growing on decaying plant materials, so its presence on raw fruits and vegetables is not rare. Surveys of fresh produce have revealed its presence on cabbage, cucumbers, potatoes and radishes, ready-to-eat salads, tomatoes and cucumbers, and bean sprouts, sliced cucumbers and leafy vegetables.

A cabbage-associated outbreak of listeriosis has been documented. Listeria monocytogenes grows at temperatures as low as 2°C and can thrive in cool, wet areas in processing facilities. It can also grow on endive, lettuce, tomatoes, asparagus, broccoli and cauliflower and cabbage but appears to be inhibited by carrot juice. Production of 6-methoxymellein and 6-hydroxymellein by carrot cells infected by fungi or upon partial hydrolysis is known to occur. These phytoalexins inhibit a wide range of spoilage and pathogenic bacteria.

The mechanism of action of 6-methoxymellein apparently involves interference with membrane-associated functions. Controlled atmosphere storage has been shown to extend the shelf-life of broccoli and asparagus but does not influence the rate of growth of L. monocytogenes). The risk of listeriosis increases when these vegetables are stored for longer periods before consumption because L. monocytogenes has a greater opportunity to increase.

Staphylococcus aureus

Staphylococcus aureus is known to be carried in the nasal passages of healthy food handlers and has been detected on raw produce and ready-to-eat vegetable salads. However, enterotoxigenic S. aureus does not compete well with other microorganisms normally present on raw fruits and vegetables, so spoilage caused by nonpathogenic microflora would probably precede the development of the high population of this pathogen that would be needed for production of staphylococcal enterotoxin.

Clostridium species

Spores of Clostridium botulinum and Clostridium perfringens can be found both in soil and on raw fruits and vegetables. The high rate of respiration of salad vegetables can create an anaerobic environment in film-wrapped packages, thus favouring the growth of C. botulinum and botulinal toxin production. Botulism has been linked to coleslaw prepared from packaged, shredded cabbage and chopped garlic in oil. Studies have revealed that C. botulinum can produce toxin in polyvinyl film-packaged and vacuum-packaged mushrooms. It is important that the permeability characteristics of packaging films minimise the possibility of development of anaerobic conditions suitable for outgrowth of clostridial spores.

Recognising that anaerobic pockets may develop in tightly packed produce, even when films have high rates of oxygen and carbon dioxide permeability, an additional measure to prevent growth of C. botulinum is to store produce at less than 3°C.

Bacillus Cereus

Spores of enterotoxigenic strains of Bacillus cereus are common in most types of soil. Some strains can grow at refrigeration temperatures. Foods other than raw fruits and vegetables are generally linked to illness implicating B. cereus. Illness associated with eating contaminated soy, mustard and cress sprouts has, however, been documented. Human illness tends to be restricted to self-limiting diarrhoea (enterotoxin) or vomiting (emetic toxin). However, emetic toxin-producing strains have produced liver failure and death by the foodborne route.

Vibrio Species

Vibrio species are generally the predominant bacterial species in estuarine waters and are therefore associated with a great variety of fish and seafoods. There are 12 human pathogenic Vibrio species, of which Vibrio cholerae, V. parahaemolyticus and V. vulnificus are of greatest concern. Vibrio cholerae is the causative agent of cholera, one of the few foodborne diseases with epidemic and pandemic potential. Carriage of the organism by infected humans is important in transmission of disease. Water can become contaminated by raw sewage.

Ingestion of water containing V. cholerae or of foods that are washed with contaminated water but not disinfected can lead to widespread transmission of cholera. An outbreak of cholera linked to the consumption of unpasteurised coconut milk has been documented.

Vibrio parahaemolyticus is perhaps the best described of the pathogenic vibrios in terms of its involvement in foodborne illness. Outbreaks are generally associated with consumption of contaminated raw or undercooked seafoods and are seasonal, peaking in summer when V. parahaemolyticus is at its highest population in estuarine water. Cross-contamination of raw fruits and vegetables with seafoods during handling, particular at retail locations, represents a potential mode of transmission to humans.

Of all the vibrios, V. vulnificus infection from contaminated seafoods results in the highest fatality rate, typically 60% in individuals suffering from liver diseases. As with V. cholerae and V. parahaemolyticus, it is possible that cross-contamination of raw fruits and vegetables with V. vulnificus could occur, resulting in human infection when they are consumed.

Viruses

Viruses can be excreted in large numbers by infected individuals. Although viruses will not grow in or on foods, raw fruits and vegetables may serve as vehicles for infection. Many foodassociated outbreaks of hepatitis A have been recorded. In most instances, these outbreaks have not appeared to depend on the stability of the virus in the food.

Shellfish taken from waters contaminated with human faeces have been the vehicle in most outbreaks, but any food handled by an infected person may become contaminated and transmit infection. Hepatitis A infection has been linked to the consumption of lettuce, diced tomatoes raspberries and strawberries. Hernandez et al. suggested that lettuce contaminated with sewage could be a vehicle for hepatitis A virus and rotavirus. Lettuce obtained from farmers' markets was reported to contain hepatitis A virus. The extent to which hepatitis A and other viruses are removed from the surface of fruits and vegetables upon treatment with chemical disinfectants is not known.

The number of cases of foodborne disease caused by Norwalklike viruses appears to be on the increase. Outbreaks have a pattern of transmission resembling that of hepatitis A. Ice made from contaminated water has been implicated as the vehicle in more than one outbreak but salad items have also been linked to Norwalk-like gastroenteritis. Workers who have prepared salads linked to viral gastroenteritis have been shown to have high antibody titers to Norwalk virus.

A non-typical outbreak of Norwalk virus gastroenteritis associated with exposure of celery to nonpotable water has been reported. Studies have shown that viruses may persist for weeks or even months on vegetable crops and in soils that have been irrigated or fertilised with sewage wastes.

Rotaviruses, astroviruses, enteroviruses (polioviruses, echoviruses and coxsackie viruses), parvoviruses, adenoviruses and coronaviruses have been reported to be transmitted by foods on occasion. At least one echovirus outbreak has been attributed to contaminated raw shredded cabbage.

Parasites

The major modes of transmission of protozoa include consumption of surface water, exposure to contaminated recreational water, animal-to-person contact and person-toperson contact. However, the epidemiology of protozoa most commonly associated with human infections, namely Giardia, Entamoeba, Toxoplasma, Sarcocystis, Isopora, Cryptosporidium, Eimeria and Cyclospora, is not fully understood.

While the life cycles of each of these parasites differ, all require passage through an animal or human host. Shedding of cysts or spores into faeces which may then, directly or indirectly, contaminate raw fruits and vegetables occurs on a global scale; it may be common in countries where hygienic conditions (especially water quality) are compromised. Expanding poverty, urban migration, food exportation, air travel and immigrant population in conjunction with other societal and environmental changes make the distinction between so-called tropical and western parasitic diseases less clear-cut than in the past.

Outbreaks of protozoan infections in humans have been linked to raw fruits and vegetables. Epidemiological evidence has implicated an asymptomatic food handler as the probable source of Giardia lamblia and raw sliced vegetables as the vehicle of transmission in an outbreak of giardiasis. Raspberries, lettuce, and basil have been the implicated vehicles of transmission in outbreaks of Cyclospora cayetanensis infection, and unpasteurised apple juice has been linked to outbreaks of cryptosporidiosis caused by Cryptosporidium parvum. A survey of vegetables has revealed the presence of Cryptosporidium oocysts on cilantro, lettuce, radish, tomato, cucumber and carrot.

The current status of knowledge about Cryptosporidium and its significance in food and beverage production has been reviewed by Donnelly and Stentiford. The presence of these protozoa on raw fruits and vegetables is likely to be due to contact with animal or human faeces, sewage, water containing untreated sewage and sludge from primary or secondary municipal water treatment facilities. While Cryptosporidium, Giardia and other parasites in water are quite resistant to chlorine and other disinfectants, little is known about the efficacy of these disinfectants in killing or removing parasites from the surface or tissues of fruits and vegetables. Surveys have shown that there is a high incidence of the parasitic roundworm, Ascraris, in the sewage sludge of many cities.

Foodborne trematode infections are major health problems, with an estimated 40 million persons, mainly in eastern and southern Asia, being affected. Infection takes place through the consumption of raw plants, or raw or undercooked freshwater fish or shellfish, containing the infective cyst (metacercaria) stage of these parasites. Watercress is a major source of Fasciola hepatica infection Over 300,000 clinical cases of fascioliasis may have occurred in more than 55 countries in Africa, the Americas, Asia, Europe and the western Pacific from 1970 to 1990. Large endemic areas have been reported more recently in Bolivia, Egypt, Iran and Peru.

A wide range of other aquatic plants may support metacercariae. Conditions for transmission of Fasciolopsis buski are present in areas of cultivation of water caltrop, water chestnut, water hyacinth, water bamboo, water mimosa, lotus and duckweed. If animal manure or effluent from livestock pens or abattoirs is used as fertiliser for these plants, it introduces Fasciolopsis to the aquatic environment. Plantborne trematodes encyst as metacerceriae on the surface of plants or on debris floating on the water surface. Plants that grow in water are also believed to serve as hosts for encystment of metacercariae of certain intestinal flukes. Nonaquatic plants such as lettuce, alfalfa, mint and sugarcane which may be eaten raw have also been implicated in human trematode infections.

Strategies for controlling foodborne trematode infections associated with eating raw plant materials will be shaped by ecological and environmental factors. The use of noncomposted animal manures or effluent from livestock pens for the purpose of fertilising aquatic plants should cease. Freesing, heating and irradiation can be effective in killing trematodes. The effectiveness of chemical treatments in killing trematodes attached to plant surfaces is not well understood.

Mitigating the risk of disease

To minimise the risk of infection or intoxication associated with raw fruits and vegetables, potential sources of contamination from the environment to the table should be identified and specific measures and interventions to prevent and/or minimise the risk of contamination should be considered and correctly implemented. Where the possibility of contamination cannot be excluded, the application of the most effective decontamination processes should be considered.

Application of good hygienic practice during production, transport and processing, combined with the Hazard Analysis Critical Control Point system, will certainly minimize the contamination of fruits and vegetables and reduce the risk of illness associated with these foods. However, food handlers and consumers also need to observe good hygienic practice during processing and preparation of these foods for consumption -including treatments for reducing the number of pathogens.

The simple practice of washing raw fruits and vegetables in hot water or water containing detergent or permanganate salts removes a portion of the pathogenic and spoilage microorganisms that may be present, but studies showing the efficacy of these treatments are few. Even washing fruits and vegetables in potable water, then again washing or rinsing in potable water would aide in removing microorganisms.

Spoilage of Meat Products

Staphylococci, streptococci, Clostridium, and Salmonella have been isolated from the lymph nodes of red-meat animals. Normal slaughtering practices would remove the lymph nodes from edible parts. The important contamination however, comes from external sources during bleeding, handling, and processing. During bleeding, skinning, and cutting, the main sources of microorganisms are the exterior of the animal and the intestinal tract. Recently approved "humane" methods of slaughter-mechanical, chemical, and electrical-have little effect on contamination, but each method is followed by sticking and bleeding, which can introduce contamination.

As with the older methods of use of a knife on hogs and poultry, any contaminating bacteria on the knife soon will be found in meat in various parts of the carcass, carried there by blood and lymph. The exterior of the animal harbors large numbers and many kinds of microorganisms from soil, water, feed, and manure, as well as its natural surface flora, and the intestinal contents contain the intestinal organisms. Knives, cloths, air, and hands and clothing of the workers can serve as intermediate sources of contaminants.

During the handling of the meat thereafter, contamination can come from carts, boxes, or other containers; other contaminated meat; air; and personnel. Especially undesirable is the addition of psychrotrophic bacteria from any source, e.g., from other meats that have been in chilling storage. Special equipment such as grinders, sausage stuffers and casings, and ingredients in special products, e.g., filters and spices, may add undesirable organisms in appreciable numbers. Growth of microorganisms on surfaces touching the meats and on the meats themselves increases their numbers.

Because of the varied sources, the kinds of microorganisms likely to contaminate meats are many. Molds of many genera may reach the surfaces of meats and grow there. Especially important are species of the genera Cladosporium, Sportrichum, Geotrichum, Thamnidium, Mucor, Penicillium, Alternaria, and Monilia. Yeasts, mostly asporogenous ones, often are present. Bacteria of many genera are found, among which some of the more important are Pseudomonas, Acinetobacter, Moraxella, Alcaligenes, Micrococcus.

Strptococcus, Sarcina, Leuconostoc, Lactobacillus, Proteus, Flavobacterium, Bacillus, Clostridium, Escherichia, Campylobacter, Salmonella, and Streptomyces. Many of these bacteria can grow at chilling temperatures. There also is the possibility of the contamination of meat and meat products with human pathogens, especially those of the intestinal type. In the retail market and in the home additional contamination usually takes place. In the market knives, saws, cleavers, slicers, grinders, chopping blocks, scales, sawdust, and containers as well as the market operators may be sources of organisms. In the home refrigerator containers used previously to store meats can serve as sources of spoilage organisms.

Raw meat is subject to change by its own enzymes and by microbial action, and its fat may be oxidised chemically. A moderate amount of autolysis is desired in the tenderising of beef and game by hanging, or aging, but is not encouraged in most other raw meats. Autolytic changes include some proteolytic action on muscle and connective tissues and slight hydrolysis of fats. The defect caused by excessive autolysis has been called "souring," an inexact term that is applied to a variety of kinds of spoilage of food and to in fact almost any kind that gives a sour odour.

Souring due to autolysis is difficult to separate or distinguish from defects caused by microbial action, especially from simple proteolysis. However, this preliminary hydrolysis of proteins by the meat enzymes undoubtedly helps microorganisms start growing in the meat by furnishing the simpler nitrogen compounds needed by many microorganisms that cannot attack complete native proteins.

General Principles

It has been pointed out that during slaughter, dressing, and cutting, microorganisms come chiefly from the exterior of the animal and its intestinal tract but that more are added from knives, cloths, air, workers, carts, boxes, and equipment in general. A great variety of kinds of organisms are added, and so it can be assumed that under ordinary conditions most kinds of potential spoilage organisms are present and will be able to grow if favourable conditions present themselves.

Invasion of Tissues by Microorganisms

Upon the death of the animal, invasion of the tissues by contaminating microorganisms takes place. Factors that influence that invasion include the following:

The load in the gut of the animal: The greater the load, the greater the invasion of tissues. For that reason starvation for 24 hr before slaughter has been recommended. The physiological condition of the animal immediately before slaughter. If the animal is excited, feverish, or fatigued, bacteria are more likely to enter the tissues, bleeding is apt to be incomplete, thus encouraging the spread of bacteria, and chemical changes may take place more readily in the tissue, such as those due to better bacterial changes may take place more readily in the tissue, such as those due to better bacterial growth because of a higher pH, earlier release of juices from the meat fibers, and more rapid denaturation of proteins. Because glycogen is used up in fatigue, the pH will not drop from 7.2 to about 5.7, as it would normally.

The method of killing and bleeding: The better and more sanitary the bleeding, the better the keeping quality of the meat. Little has been reported on the effect of humane methods of slaughter on the keeping quality of meat, although it has been claimed that more greening was found in pork and bacon from electrically stunned animals than from those killed with carbon dioxide.

The rate of cooling: Rapid cooling will reduce the rate of invasion of the tissues by microorganisms. Microorganisms are spread in the meat through the blood and lymph vessels and connective-tissue interspaces, and in ground meat by grinding.

Growth of Microorganisms

Meat is an ideal culture medium for many organisms because it is high in moisture, rich in nitrogenous foods of various degrees of complexity, and plentifully supplied with minerals and accessory growth factors. Also, it usually has some fermentable carbohydrate and is at a favourable pH for most microorganisms. The factors that influence the growth of microorganisms are briefly explained as follows.

The Kind and Amount of Contamination with Microorganisms and the Spread of these Organisms in the Meat: For example, meat with a contaminating flora is high in percentage of psychrotrophs would spoil at chilling temperatures more rapidly than meat with a low percentage of these psychrotrophic organisms.

Physical Properties of the Meat: The amount of exposed surface of the flesh has considerable influence on the rate of spoilage because the greatest load of organisms usually is there and air is available for aerobic organisms. Fat may protect some surfaces but is subject to spoilage itself, chiefly enzymatic and chemical. Grinding meat greatly increases the surface and encourages microbial growth for this reason and because it releases moisture and distributes bacteria throughout the meat. Skin on meat serves to protect the meat inside, although microorganisms grow on it.

Chemical Properties of the Meat: It has been pointed out that meat in general is a fine culture medium for microorganisms. The moisture content is important in determining whether organisms can grow and what kinds can grow, especially at the surface, where drying may take place. Thus the surface may be so dry as to permit no growth, a little moist to allow mold growth, still moister to encourage yeasts, and very moist to favour bacterial growth. The relative humidity of the storage atmosphere is important in this regard.

Food for microorganisms is plentiful, but the low content or absence of fermentable carbohydrate and the high protein content tend to favour the nonfermenting types of organisms, those which can utilise proteins and their decomposition products for nitrogen, carbon, and energy. The pH of raw meat may vary from about 5.7 to over 7.2, depending on the amount of glycogen present at slaughter and subsequent changes in the meat. A higher pH value favours microbial growth; a lower one usually makes it slower and may be selective for certain organisms, such as yeasts.

Availability of Oxygen: Aerobic conditions at the surface of meat are favourable to molds, yeasts, and aerobic bacteria. Within the solid pieces of meat, conditions are anaerobic and tend to remain that way because the potential is strongly poised at a low level, although oxygen will diffuse slowly into ground meat and slowly raise the O-R potential unless the casing or packaging material is impervious to oxygen. True putrefaction is favoured by anaerobic conditions.

Temperature: Meat should be stored at temperatures not far above freezing, where only low-temperature microorganisms can grow. Molds, yeasts, and psychrotrophic bacteria grow slowly and produce characteristic defects to be discussed later. True putrefaction is rare at these low temperatures but is likely at room temperature. As for most foods, the temperature is most important in selecting the kinds of organisms to grow and the types of spoilage to result.

At chilling temperatures, for example, psychrophiles are favoured and proteolysis is likely, caused by a dominating species of bacterium, followed by utilisation of peptides and amino acids by secondary species. At ordinary atmospheric temperatures, mesophiles would grow, such as coliform bacteria and species of Bacillus and Clostridium, with the production of moderate amounts of acid from the limited amounts of carbohydrates present.

General Types of Spoilage

The common types of spoilage of meats can be classified on the basis of whether they occur under aerobic or anaerobic conditions and whether they are caused by bacteria, yeasts, or molds.

Under Aerobic Conditions

Under aerobic conditions bacteria may cause the following:

— *Surface slime*, which may be caused by species of Pseudomonas, Acinetobacter, Moraxella, Alcaligenes, Streptococcus, Leuconostoc, Bacillus, and Micrococcus. Some species of Lactobacillus can produce slime. The temperature and the availability of moisture influence the kind of microorganisms causing surface slime. At chilling temperatures, high moisture will favour the Pseudomonas-Alcaligenes group; with less moisture, as on frankfurters, micrococci and yeasts will be encouraged; and with still less moisture, molds may grow. At higher temperatures, up to that of the room, micrococci and other mesophiles compete well with the pseudomonads and related bacteria. The numbers of microorganisms necessary before detection of off-odour or slime in meats. Numbers in the millions per square centimeter or gram are required.

— *Changes in color of meat pigments*: The red color of meat, called its "bloom," may be changed to shades of green, brown, or gray as a result of the production of oxidising compounds, e.g., peroxides, or of hydrogen sulfide, by bacteria. Species of Lactobacillus and Leuconostoc are reported to cause the greening of sausage.

— *Changes in fats*: The oxidation of unsaturated fats in meats takes place chemically in air and may be catalysed by light and copper. Lipolytic bacteria may cause some lipolysis and also may accelerate the oxidation of the fats. Some fats, such as butterfat, become tallowy on oxidation and rancid on hydrolysis, but most animal fats develop oxidative rancidity when oxidised, with off-odours due to aldehydes and acids. Hydrolysis adds the flavour of the released fatty acids. Racidity of fats may be caused by lipolytic species of Pseudomonas and Achromobacter or by yeasts.

— *Phosphorescence*: This rather uncommon defect is caused by phosphorescent or luminous bacteria, e.g., Photobacterium spp., growing on the surface of the meat.

— *Various surface colors due to pigmented bacteria*: Thus "red spot" may be caused by Serratia marcescens or other bacteria with red pigments. Pseudomonas syncyanea can impart a blue color to the surface. Yellow discolorations are caused by bacteria with yellow pigments, usually species of Micrococcus of Flavobacterium. Chromobacterium lividum and other bacteria give greenish-blue to brownish-black spots on stored beef. The purple "stamping-ink" discoloration of surface fat is caused by yellow-pigmented cocci and rods. When the fat becomes rancid and peroxides appear, the yellow color changes to a greenish shade and later becomes purplish to blue.

— *Off-odours and off-tastes*: "Taints," or undesirable odours and tastes, that appear in meat as a result of the growth of bacteria on the surface often are evident before other signs of spoilage. "Souring" is the term applied to almost any defect that gives a sour odour that may be due to volatile acids, e.g., formic, acetic, butyric, and propionic, or even to growth of yeasts. "Cold-storage flavour" or taint is an indefinite term for a stale flavour. Actinomycetes my be responsible for a musty or earthy flavour.

Under aerobic conditions yeasts may grow on the surface of meats, causing sliminess, lipolysis, off-odours and tastes, and discolorations-white, cream, pink, or brown-due to pigments in the yeasts.

Aerobic growth of molds may cause the following:

— *Stickiness*: Incipient growth of molds makes the surface of the meat sticky to the touch.

— *Whiskers*. When meat is stored at temperatures near freezing, a limited amount of mycelial growth may take place without sporulation. Such white, fuzzy growth can be caused by a number of molds, including Thamnidium chaetocladioides, or T. elegans; Mucor mucedo, M. lusitanicus, or M. racemosus; Rhizopus; and others. Controlled growth of a special strain of Thamnidium has been recommended for improvement in flavour during aging of beef.

— *Black spot*: This usually is caused by Cladosporium herbarum, but other molds with dark pigments may be responsible.

— *White spot*: Sporotrichum carnis is the most common cause of white spot, although any mold with wet, yeastlike colonies, e.g., Geotrichum, can cause white spot.

— *Green patches*: These are caused for the most part by the green spores of species of Penicillium such as P. expansum, P. asperulum, and P. oxalicum.

— *Decomposition of fats*: Many molds have lipases and hence cause hydrolysis of fats. Molds also aid in the oxidation of fats.

— *Off-odours and off-tastes*: Molds give a musty flavour to meat in the vicinity of their growth. Sometimes the defect is given a name indicating the cause, e.g., "thamnidium taint."

Sports of surface spoilage by yeasts and molds usually are localised to a great extent and can be trimmed off without harm to the rest of the meat. The time that has been allowed for diffusion of the products of decomposition into the meat and the rate of that diffusion will determine the depth to which the defect will appear. Extensive bacterial growth over the surface may bring fairly deep penetration. Then, too, facultative bacteria may grow inward slowly

Spoilage under Anaerobic Conditions

Facultative and anaerobic bacteria are able to grow within the meat under anaerobic conditions and cause spoilage. The terminology used in connection with this spoilage is inexact. Most used are the words "souring," "putrefaction," and "taint," but these terms apparently mean different things to different people.

— *Souring*: The term implies a sour odour and perhaps taste. This could be caused by formic, acetic, butyric, propionic, and higher fatty acids or other organic acids such as lactic or succinic. Souring can result from

  (a) action of the meat's own enzymes during aging or ripening,

  (b) anaerobic production of fatty acids or lactic acid by bacterial action, or

  (c) proteolysis without putrefaction, caused by facultative or anaerobic bacteria and sometimes called "stinking sour fermentation." Acid and gas formation accompany the action of the "butyric" Clostridium species and the coliform bacteria on carbohydrates. Vacuum-packed meats, especially those in gastight wrappers, commonly support the growth of lactic acid bacteria.

— *Putrefaction*: True putrefaction is the anaerobic decomposition of protein with the production of foulsmelling compounds such as hydrogen sulfide, mercaptants, indole, skatole, ammonia, and amines. It usually is caused by species of Clostridium, but facultative bacteria may cause putrefaction or assist in its production, as evidenced by the long list of species with the specific names putrefaciens, putrificum, putida, etc., chiefly in the genera Pseudomonas and Alcaligencs. Also, some species of Proteus are putrefactive. The confusion in the use of the term "putrefaction" arises from the fact that any type of spoilage with foul odours, whether from the anaerobic decomposition of protein or from the breakdown of other compounds, even nonintrogenous ones, may erroneously be termed putrefaction. Thus, for example, trimethylamine in fish and isovaleric acid in butter are described as putrid odours. Gas formation accompanies putrefaction by clostridia, the gases being hydrogen ad carbon dioxide.

— *Taint*: "Taint" is a still more inexact word applied to any offtaste or off-odour. The term "bone taint" of meats refers to either souring or putrefaction next to the bones, especially in hams. Usually it means putrefaction.

Not only air but temperature has an important influence on the type of spoilage to be expected in meat. When meat is held at temperatures near 0 C, as recommended, microbial growth is limited to that of molds, yeasts, and bacteria able to growth at low temperatures. These include many of the types that produce sliminess, discoloration, and spots of growth on the surface and many that can cause souring, such as Pseudomonas, Acinetobacter Moraxella, Alcaligenes, Lactobacillus, Leuconostoc, Streptococcus, and Flavobacterium species. Most true putrefiers, like those in the genus Clostridium, require temperatures above those of the refrigerator.

Spoilage of Milk

Milk is sterile at secretion in the udder but is contaminated by bacteria even before it leaves the udder. Except in the case of mastisis, the bacteria at this point are harmless and few in number. Further infection of the milk by microorganisms can take place during milking, handling, storage, and other preprocessing activities.

Lactic acid bacteria: this group of bacteria are able to ferment lactose to lactic acid. They are normally present in the milk and are also used as starter cultures in the production of cultured dairy products such as yogurt.

Coliforms: coliforms are facultative anaerobes with an optimum growth at 37° C. Coliforms are indicator organisms; they are closely associated with the presence of pathogens but not necessarily pathogenic themselves. They also can cause rapid spoilage of milk because they are able to ferment lactose with the production of acid and gas, and are able to degrade milk proteins. They are killed by HTST treatment, therefore, their presence after treatment is indicative of contamination. Escherichia coli is an example belonging to this group.

Information on the microbial content of milk can be used to judge its sanitary quality and the conditions of production. If permitted to multiply, bacteria in milk can cause spoilage of the product Milk is potentially susceptible to contamination with pathogenic microorganisms. Precautions must be taken to minimise this possibility and to destroy pathogens that may gain entrance Certain microorganisms produce chemical changes that are desirable in the production of dairy products such as cheese, yogurt.

The microbial quality of raw milk is crucial for the production of quality dairy foods. Spoilage is a term used to describe the deterioration of a foods' texture, colour, odour or flavour to the point where it is unappetising or unsuitable for human consumption. Microbial spoilage of food often involves the degradation of protein, carbohydrates, and fats by the microorganisms or their enzymes.

In milk, the microorganisms that are principally involved in spoilage are psychrotrophic organisms. Most psychrotrophs are destroyed by pasteurisation temperatures, however, some like Pseudomonas fluorescens, Pseudomonas fragi can produce proteolytic and lipolytic extracellular enzymes which are heat stable and capable of causing spoilage. Some species and strains of Bacillus, Clostridium, Cornebacterium, Arthrobacter, Lactobacillus, Microbacterium, Micrococcus, and Streptococcus can survive pasteurisation and grow at refrigeration temperatures which can cause spoilage problems.

Pathogenic Microorganisms

Hygienic milk production practices, proper handling and storage of milk, and mandatory pasteurisation has decreased the threat of milkborne diseases such as tuberculosis, brucellosis, and typhoid fever. There have been a number of foodborne illnesses resulting from the ingestion of raw milk, or dairy products made with milk that was not properly pasteurised or was poorly handled causing post-processing contamination. The following bacterial pathogens are still of concern today in raw milk and other dairy products:

- Bacillus cereus
- Listeria monocytogenes
- Yersinia enterocolitica
- Salmonella spp.
- Escherichia coli O157:H7
- Campylobacter jejuni

It should also be noted that moulds, mainly of species of Aspergillus, Fusarium, and Penicillium can grow in milk and dairy products. If the conditions permit, these moulds may produce mycotoxins which can be a health hazard.

Contamination Sources

Under normal grazing conditions, cows' udders will appear clean and therefore washing and drying will be unnecessary. Otherwise, any visible dirt must be removed using clean, running water, individual paper towels or cloths in clean water to which a disinfectant has been added. If udder cloths are used, provide a clean cloth for each cow. After each milking wash and disinfect them and hang up to dry. Disposable paper towels are preferable and more effective for drying after washing.

When cows are housed or graze in heavily stocked paddocks, external udder surfaces are usually grossly contaminated with bacteria even when they appear visibly clean, therefore routine udder preparation procedures should be followed. Whenever udders are washed they should be dried. Foremilking has little affect on the total bacterial count of the milk but is an effective way of detecting clinical symptoms of mastitis.

Filtering or straining the milk removes visible dirt but not the bacteria in the milk because they pass through the filter. Aerial contamination of milk by bacteria is insignificant under normal production conditions. The milk contact surfaces of milking and cooling equipment are a main source of milk contamination and frequently the principal cause of consistently high bacterial counts. Simple, inexpensive cleaning and disinfecting routines can virtually eliminate this source of contamination.

Bacterial Concern in Milk

Psychrotrophic Bacteria

Psychrotrophs are bacteria which are capable of growing at 7°C (44.6°F) or less. Psychrotrophs are of primary concern to the dairy industry since they can grow and cause spoilage in raw or processed dairy products commonly held under refrigeration.

— The most commonly occurring psychrotrophs in milk are Gram-negative rods, mostly belonging to the genus Pseudomonas. Gram-negative psychrotrophs generally do not survive pasteurisation thus they occur in processed milk as post-pasteurisation contaminants.

— Psychrotrophs may be present in low numbers in untreated water supplies which are used for rinsing and washing dairy equipment. Dirty equipment and marginal cooling can result in relatively large numbers of these organisms in milk.

— Psychrotrophic bacteria produce a variety of enzymes that cause chemical deterioration of milk resulting in off-flavors. Some of these enzymes are not inactivated by pasteurisation or by other heat treatments and may continue to degrade milk products, even when the bacterium is destroyed.

Thermoduric Bacteria

Thermoduric bacteria are a miscellaneous group of bacteria that are capable of surviving pasteurisation or other heat treatments.

— Chief sources of thermodurics in milk are persistently poorly cleaned equipment including old rubber parts, areas of milkstone build-up, separators and other difficult to clean areas. They may contaminate milk at the farm or at the plant.

— High thermoduric counts in raw supplies may result in counts in pasteurised product that exceed legal limits (20,000/ml).

— Thermodurics are seldom psychrotrophic. However, in the absence of Gram-negative psychrotrophs certain thermoduric bacteria may grow and cause spoilage of pasteurised milk. Heat Resistant Spore-forming Psychrotrophs belonging to the genus Bacillus are considered common thermoduric psychrotrophs in some areas.

Coliform Bacteria

Coliform bacteria are by definition "aerobic or facultatively anaerobic, Gram-negative rods, that ferment lactose with the production of acid and gas". These characteristics allow selective counting of these types of bacteria.

— They are called Coliforms because some members of the group are found in the intestines (colon) of warm blooded animals. However, certain coliform bacteria are common in other natural habitats.

— Coliforms are almost always found in raw milk though with good production methods the numbers can be kept very low.

— Coliforms do not survive pasteurisation. When detected in processed milk, they indicate recontamination after pasteurisation (Post-Pasteurisation Contaminants).

Pathogenic Bacteria

Pasteurisation was originally designed to destroy pathogenic bacteria that caused tuberculosis, brucellosis and typhoid, illnesses that were often associated with the consumption of raw milk. Pasteurisation of milk has virtually eliminated these types of illnesses. Raw milk may also harbour other organisms associated with foodborne illness, including Salmonella, Listeria, Campylobacter, Yersinia and certain strains of E. coli; these organisms are also eliminated by pasteurisation. However, cross-contamination of processed dairy products with raw milk and/or the direct consumption of raw milk have resulted in relatively recent outbreaks of foodborne illnesses involving these organisms.

Bacteria in Raw Milk and Processed Milk

Milk, when synthesized in the udder of a cow, is virtually sterile. As milk passes through the teat cistern and teat channel of a healthy cow, it may be contaminated with low levels of bacteria (<1000/ml) which are generally not significant. Milk from a cow with mastitis (infection of the mammary gland) however, may harbour large numbers of the infectious bacteria. After it leaves the cow, milk may be contaminated from the exterior of the cow (dirty cows), the environment and poorly cleaned equipment.

Poor cooling can result in rapid increases in bacterial numbers in raw milk before it is processed. While the legal limit for bacteria in raw milk is 100,000/ml, the production of milk with bacteria counts less than 10,000/ml should be easily achievable for most farms. Pasteurisation, while designed to destroy potential pathogens in raw milk, substantially reduces the total numbers of bacteria present, increasing the shelf-life of the milk. Unless gross recontamination has occurred, bacterial numbers in fresh pasteurised milk generally reflect the organisms that survive pasteurisation.

The legal limit for bacterial numbers in pasteurised milk is 20,000/ml, though bacteria counts for most fresh pasteurised milk are generally less than 1,000/ml. Under proper refrigeration, the bacteria which become significant in the shelf-life and spoilage of milk are psychrotrophic in nature. These types of organisms generally occur as post-pasteurisation contaminants, though a few thermoduric bacteria may be psychrotrophs.

Sources of Bacteria

- Survive pasteurisation (thermoduric).
- Post-Pasteurisation Contamination:
    - Insufficient cleaning/sanitising—valves, pipelines, gaskets, pasteurised milk tanks, fillers.
    - Personnel—hands, sneezes, coughs.
    - Environmental—air, dust, water, condensate.

Controlling Bacteria Defects in Milk

Prevention of Contamination

Bacteria are present in the environment both at the farm and at the dairy plant. Total prevention of microbial contamination of milk at the farm is impossible, though it can be minimised by milking clean, healthy cows, in a clean environment and by assuring that the milking system and storage equipment is properly cleaned, sanitised and maintained. Once raw milk leaves the farm it must be properly handled to prevent further contamination before it is processed. Keeping the microbial load of raw milk to a minimum will increase the quality of the products made.

At the dairy plant, preventing contamination after pasteurisation is critical for product shelf-life and safety. This requires that the processing equipment and the plant environment be thoroughly cleaned and sanitised in a manner that the possibility of microbial growth and contamination is limited. Once cleaned and sanitised, recontamination should be prevented. Proper employee training in dairy and personal hygiene procedures should be a part of every plant's quality assurance programme.

Slowing Microbial Growth

Microbial growth can be controlled by:

1) eliminating sources of "bacterial food" by thoroughly cleaning the milk handling equipment and the environment, thus eliminating milk residues and other sources of microbial nutrition, at the farm, during transit and at the plant;
2) holding raw milk and dairy foods well below the optimum growth temperature of bacterial contaminants, generally less than 40°F without freezing;
3) lowering the pH such as in cultured dairy products;
4) reducing the moisture or water activity (Aw) such as in dry milk products; and
5) adding microbial inhibitors or preservatives such as is done with potassium sorbate addition to cottage cheese.

Sanitation Procedures

Chemical sanitisers are routinely used to reduce the load of microbial contaminants that may be present on milk/food contact surfaces. Most dairy sanitisers, when used correctly, kill off a broad spectrum of microorganisms. Sanitation procedures should be performed after washing and immediately before processing, though an additional sanitising step after equipment washing procedures can be helpful. Most chemical sanitisers are inactivated by organic matter and are ineffective on poorly cleaned surfaces.

Sanitisers commonly used in the dairy industry include chlorine and iodine compounds, quaternary ammonium compounds, acid anionics and peroxyacetic acid. Hot water sanitation is commonly used in many dairy plants. Hot water sanitation involves circulating water of at least 170°F (determined at the outlet) for at least 5 minutes. Higher temperatures (>185°F) for longer times (10-15 minutes) are recommended to allow heat penetration into areas that are hard to reach. Hot water treatments should be followed by a cooling chemical sanitiser rinse. Hot water will often provide greater kill and longer milk shelf-life than can be achieved with chemical sanitisers alone.

Control of Spoilage Microorganisms

Spoilage organisms are not originally an integral part of foods but are widely present in water, soil, air, and other animals. Healthy living plants and animals can ward off bacteria and fungi, but as soon as they are slaughtered or harvested their defenses deteriorate and their tissues become susceptible to spoilage microbes. Good manufacturing practices with strict attention to sanitation and hygiene can prevent colonization by many, but not all, microbes and are the most important first step in delaying the spoilage process.

Microbes require certain conditions for growth, and therefore management of the environment of foods can change these factors and delay spoilage:

— Many, but not all, microbes grow slowly or not at all at low temperatures, and refrigeration can prolong the lag phase and decrease growth rate of microbes.

— Many microbes require a high water activity and therefore keeping foods such as grains and cereal products dry will help to preserve them.

— Some microbes require oxygen, others are killed by oxygen, and still others are facultative. Managing the atmosphere during storage in packaging can retard or prevent the growth of some microbes. Several types of modified atmosphere packaging (MAP) have been developed to retard growth of pathogenic and spoilage *organisms.*

However, microbes are endlessly innovative and eventually seem to circumvent the barriers we set up against them. Therefore further strategies and multiple hurdles are utilized to extend shelf life. These procedures must be assessed for compatibility with different foods so that there are no significant organoleptic changes in the foods caused by the treatment or preservative. These methods for food preservation will not be covered in depth here.

Processing technologies, in addition to thermal processing, are being developed to kill spoilage microbes, including:

— high pressure processing of fruits, juices, meat and fish

— ozone

— irradiation of fruit and meat

— pulsed electric fields of juices

Formulation of processed foods may include compounds that alter the water activity or pH of foods, thereby limiting growth of many organisms.

Antimicrobial compounds may be added to foods or packaging to inhibit growth of many spoilage organisms:

— Organic acids can help control bacteria, molds and yeasts in bakery products, meat, juices, and other foods

— Bacteriocins, including nisin, can help control spoilage bacteria in dairy products, fish, juice, and vegetables

— Chitosan incorporated into foods or used as a coating for fruits and vegetables inhibits growth of some spoilage bacteria and yeasts

— Many herbs, essential oils, and spices have demonstrated some inhibitory activity against spoilage microbes in a variety of foods. Thyme, oregano, vanillin, and cinnamon are the most commonly mentioned substances in recent papers

References

Davidson, Alan.,2006. *The Oxford Companion to Food,* 2nd ed, UK: Oxford University Press,

Gram L and Dalgaard P. 2002. Fish spoilage bacteria—problems and solutions. *Curr Opin Biotechnol.* 13:262–266.

Hui, Y.H., 1994. *Foodborne Disease Handbook,* M. Dekker.

Loureiro V and Malfeito-Ferreira M. 2003. Spoilage yeasts in the wine industry. Int *J Food Microbiol* 86:23–50.

Pitt JI and Hocking AD. 1997. *Fungi and Food Spoilage*. Blackie Academic and Professional, New York.

Smits GJ and Brul S. 2005. Stress tolerance in fungi—to kill a spoilage yeast. *Curr Opin Biotechnol* 16:225–230.

Tournas VH. 2005. Spoilage of vegetable crops by bacteria and fungi and related health hazards. *Crit Rev Microbiol* 31:33–44.

5

Mycotoxins and Food Contamination

The experience of the past 25 years indicates that it is possible to maintain present food consumption levels by increasing overall food supplies in quantitative terms. However, in terms of providing food of the right quality which is nutritious and free from environmental contaminants, the task ahead is challenging, particularly in highly populated parts of the world. Among food contaminants, mycotoxins will have greater consequences in terms of both human and animal health as well as economics.

Mycotoxins are substances produced by moulds that contaminate various agricultural commodities either before harvest or under post-harvest conditions. In addition to the various moulds occurring in crops which are improperly stored, certain plant diseases are responsible for the production of mycotoxins. Different weather conditions, such as unseasonable rains at the time of flowering or cyclones and droughts during harvest and post-harvest stages; mould growth; and mycotoxin contamination can also pose serious problems.

Mycotoxins have received considerable attention especially over the last three decades. Mycotoxicology is currently a subject of international importance. The problem of mould damage and the hazard of consuming damaged grains have been recognized since historical times. In areas of the Indo-Gangetic plains, for instance, traditional farmers have practised appropriate post-harvest measures to preserve crops. Expressions like "dry the grains well and keep dry grains dry" have been passed from one generation to the next. The problem of ergotism resulting from the infestation of rye by *Claviceps* has been known since biblical times and is now recognized as having been an important cause of human mortality in medieval Europe. The use of wheat, which is highly susceptible to *Fusarium*species producing trichothecenes, increased during this period and has been linked to the plague epidemics of the time. During the first half of this century, the possibility of human diseases occurring as a result of the consumption of mould-damaged rice and wheat was raised in Japan and other Asian countries. In the USSR, there was also awareness of risks from eating overwintered millet. However, the serious worldwide concern about mycotoxins began in the early 1960s after it was discovered in the United Kingdom that Turkey "X" disease is caused by aflatoxins.

Occurrence of Mycotoxins

Although there are many species of toxigenic moulds, only a few mycotoxins, particularly those affecting cereals (maize, wheat, barley, oats and rice) and groundnuts, are considered to be significant for humans. The most well-known mycotoxin, the potent human hepatocarcinogen aflatoxin, is produced by *Aspergillus flavus* and *A. parasiticus.* These moulds occur in warm climates and produce aflatoxin in drought-stressed maize and groundnuts in the field. They also affect these crops and many other commodities (copra, cottonseed, pepper) which are stored under improper conditions of temperature and humidity.

There are 24 toxigenic species of *Fusarium* which are increasingly viewed as having an important effect on human and animal health. *F. graminearum,* which causes head blight and ear rot, produces a variety of potent mycotoxins including deoxynivalenol, zearalenone and fusarin C. *F. sporotrichioides* produces T-2 toxin and related compounds which were responsible for the large-scale human toxicosis in Eastern Europe noted above.

Epidemics of *Fusarium* head blight and maize ear rot are chronic in cereal-growing areas of Asia as well as parts of Africa and South America. Public health authorities in these areas report acute human toxicosis from deoxynivalenol. The effect of chronic exposure to trichothecenes (which are immunosuppressors) is not known. Of great concern at the moment are the toxins of *F. moniliforme,* namely fusarins and fumonisins. The presence of these toxins in maize in southern Africa and parts of China may explain the high rates of oesophageal cancer in these regions. There are a number of other species of moulds, particularly *Aspergillus* and *Penicillium,* that produce mycotoxins under poor storage conditions. Ochratoxin is thought to be the most important mycotoxin in this category.

There is no information available on the precise extent of contamination and co-occurrence of various mycotoxins in different commodities on a worldwide basis. Data from the Joint UNEP/FAO/WHO Food Contamination Monitoring Programme[1] indicated that "much of the grain monitored in the United States and the Soviet Union, and to a lesser extent in Guatemala and Kenya, contained 90th percentile levels above 5-20 m g/kg, the regulatory levels for foods in most countries". Data from 185000 samples of peanuts from Brazil, Guatemala, Ireland, Mexico, Switzerland, the United Kingdom, the United States and the USSR indicated that, for all but 1 percent of the samples, the 90th percentile levels of aflatoxins were less than 20 m g/kg.

Although there is little firm evidence, the results of these surveys suggest that in certain situations as much as 50 percent of the grain may be contaminated with mycotoxins. In certain years, the deoxynivalenol content in wheat was found to be more than 500 m g/kg in some countries.

The poorest quality grain (where it can be spared) is used for animal feed. Animal feeds with ingredients such as oilseed cakes, peanut, cottonseed and coconut cake or corn grits often contain mycotoxins. Feed conversion to animal protein is always reduced by the presence of mycotoxins. In addition, mycotoxins have a negative effect on animal health and fertility is decreased. When animals ingest the contaminated feeds, some toxins can be metabolized and remain in milk, meat and eggs. The presence of mycotoxins in animal products such as aflatoxins and ochratoxins in milk, meat and eggs has been of concern in some countries.

The levels of toxins such as aflatoxin and ochratoxins present in these secondary sources are much lower than those in agricultural commodities. However, their levels in these products, in particular milk, is strictly regulated in most developed countries. The effects of this source of toxins on human health may be modest in developed countries because of feed safety regulations and pooling of milk at dairies. However, in developing countries where animals are likely to consume mycotoxin-contaminated animal feeds and are milked individually at the household level, the levels of toxins can be higher. Despite this, mycotoxin occurrence in animal products in some high-risk areas is being paid little attention.

Health Implications

The consumption of mycotoxin-contaminated commodities is related to several acute and chronic diseases in humans as well as in animals. While the exact cause and effect relationship has been established for only a few of the diseases, speculation about the role of mycotoxins in the aetiology of various illnesses has been based on circumstantial evidence in other cases. The acute diseases for which there is some evidence of an association with mycotoxins include: aflatoxic hepatitis in India and Kenya; enteric ergotism in India; vascular ergotism in Ethiopia; and deoxynivalenol mycotoxicosis in India and China. A common feature in all these outbreaks has been the involvement of staple foods such as corn, wheat or pearl millet, following unseasonable rains or drought during either the growing season or harvest.

Among the mycotoxins, aflatoxins have been implicated in human diseases including liver cancer, Reye's syndrome, Indian childhood cirrhosis, chronic gastritis, kwashiorkor and certain occupational respiratory diseases in various parts of the world, particularly in African and Asian countries. In China, the Philippines, Thailand, Kenya, Swaziland and Mozambique, higher levels of aflatoxins in the food supply have been correlated with aflatoxins and their derivatives in human fluids which may be associated with liver cancer. *Fusarium* toxins have been suspected to have a role in diseases such as Kashin Beck syndrome in the USSR, China and Viet Nam; Mseleni joint disease in southern Africa; endemic familial arthritis in India; alimentary toxic aleukia in the USSR; and oesophageal cancer in southern Africa. Ochratoxins have been associated with Balkan endemic nephropathy and urinary tract tumours. However, in most of these instances, conclusive evidence for the role of mycotoxins in disease causation has been lacking.

Mycotoxicology—A Systems Approach

A 'system' may be viewed as a set of interacting components, where the interactions are just as important as the components themselves. A 'systems' approach to the control of mycotoxins utilises conceptual models of interactions between, and within, commodity, spoilage, mycotoxin, and control subsystems. Within a system, the sub-systems can freely interact; in other words, activity within one subsystem can influence events in one or more other sub-systems.

A better understanding of both the interactions and the components associated with these systems will assist in understanding the aetiology of mycotoxin production, and in formulating appropriate interventions for the control of mycotoxins and mycotoxicoses.

The Commodity System

Any commodity system is composed of numerous interacting technical and socio-economic 'processes' including, for example, pest and disease control, harvesting, drying, processing, marketing, credit and pricing policies and cultural issues, to name but a few.

At any point within the commodity system, the condition of the commodity is determined by a complex milieu involving a multitude of interactions between the crop, the macro- and micro-environment and a variety of biological, chemical, physical and socio-economic factors. A change within any one process will invariably bring about changes in one or more of the other processes. Action taken before harvest to control pest damage and/or increase production (e.g selection of varieties, timing of harvest) can have a significant impact on the post-harvest quality of the commodity. Hybrid white maize, for example, has much higher yields than traditional varieties but has poor on-farm storage characteristics. Similarly, since it is very rare for a single commodity system to exist in isolation within a given agro-climatic region, it should be remembered that activities within one system can significantly effect events in other systems. Given the finite resources of farmers, for example, an increase in the importance of one commodity will frequently lead to the allocation of less resources towards the care of other commodities.

The Spoilage System

Biodeterioration is the net result of numerous interacting spoilage agents which may be broadly described as biological, chemical, physical, macro-environmental and micro-environmental. However, the relative impact of these agents will often be largely determined by the nature and extent of human intervention.

The factors which primarily contribute to bioterioration (including mould growth) within an ecosystem, are moisture, temperature and pests. Moulds can grow over a wide range of temperatures and, in general, the rate of mould growth will decrease with decreasing temperature and available water. In grains, moulds utilise intergranular water vapour, the concentration of which is determined by the state of the equilibrium between free water within the grain (the grain moisture content) and water in the vapour phase immediately surrounding the granular particle. The intergranular water concentration is described either in terms of the equilibrium relative humidity (ERH, %) or water activity (a_w). The latter describes the ratio of the vapour pressure of water in the grain to that of pure water at the same temperature and pressure, whilst the ERH is equivalent to the water activity expressed as a percentage. For a given moisture content, different grains afford a variety of water activities and, consequently, support differing rates and type of mould growth. Typical water activities which are necessary for mould growth range from 0.70 to 0.99, the water activity, and the propensity for mould growth increasing with temperature. Maize, for example, can be relatively safely stored for one year at a moisture level of 15 per cent and a temperature of 15°C. However, the same maize stored at 30°C will be substantially damaged by moulds within three months.

Insects and mites (arthropods) can also make a significant contribution towards the biodeterioration of grain because of the physical damage and nutrient losses caused by their activity, and also because of their complex interaction with moulds and mycotoxins. The metabolic activity of insects and mites causes an increase in both the moisture content and temperature of the infested grain. Arthropods also act as carriers of mould spores and their faecal material can be utilised as a food source by moulds. Furthermore, moulds can provide food for insects and mites but, in some case, may also act as pathogens.

Another important factor that can affect mould growth is the proportion of broken kernels in a consignment of grain. Broken kernels, caused by general handling and/or insect damage, are predisposed to mould invasion of the exposed endosperm.

Mould growth is also regulated by the proportions of oxygen, nitrogen and carbon dioxide in the inter-granular atmosphere. Many moulds will grow at very low oxygen concentrations; a halving of linear growth, for example, will only be achieved if the oxygen content is reduced to less than 0.14 per cent. Interactions between the gases and the prevailing water activity also influence mould growth.

The interactions described above, within granular ecosystems, will support the growth of a succession of micro-organisms, including toxigenic moulds, as the nutrient availability and microenvironment changes with time.

In the field, grains are predominantly contaminated by those moulds requiring high water activities (at least 0.88) for growth, whereas stored grains will support moulds which grow at lower moisture levels.

It is well recognised that the main factors which influence the production of mycotoxins are water activity and temperature. However, given the complexity of the ecosystems supporting the production of mycotoxins, the conditions under which toxigenic moulds produce mycotoxins are still poorly defined; and have recently been comprehensively reviewed.

The Mycotoxin System

The Mycotoxin System may be considered in terms of three interacting subsystems: metabolism & toxicology; health & productivity; and wealth. After exposure (by ingestion, inhalation or skin contact), the toxicity of a mycotoxin is determined by a sequence of events (metabolism) involving the administration, absorption, transformation, pharmacokinetics, molecular interactions, distribution, and excretion of the toxin and its metabolites. In turn, the toxicity of a mycotoxin will be manifested by its effect on the health and productivity of crops, humans and animals; and, these effects will influence the production of wealth associated with human endeavour and agricultural and livestock products.

Mycotoxins of World-wide Importance

Those moulds and mycotoxins which are currently considered to be of world-wide importance are shown in Table 1.

An 'important' mycotoxin will have demonstrated its capacity to have a significant impact upon human health and animal productivity in a variety of countries.

Table 1. Moulds and mycotoxins of world-wide importance

Mould species	*Mycotoxins produced*
Aspergillus parasiticus	Aflatoxins B_1, B_2, G_1, G_2
Aspergillus flavus	Aflatoxins B_1, B_2
Fusarium sporotrichioides	T-2 toxin
Fusarium graminearum	Deoxynivalenol (or nivalenol)
Fusarium moniliforme	Fumonisin B_1
Penicillium verrucosum	Ochratoxin A
Aspergillus ochraceus	Ochratoxin A

The Aflatoxins

The optimal water activity for growth of A. flavus is high (about 0.99). The maximum is at least 0.998 whereas the minimum water activity for growth has not been defined precisely. Pitt and Miscamble report a minimum of approximately 0.82. In general, production of toxins appears to be favoured by high water activity. A. flavus is reported to grow within the temperature range 10 - 43°C. The optimal growth rate occurs at a little above 30°C, reaching as much as 25 mm per day. The aflatoxins are produced by A. flavus over the temperature range 15 - 37°C, at least. It is not possible to specify an optimum temperature for the production of the toxins, although production between 20 - 30°C is reported to be significantly greater than at higher and lower temperatures.

The effect of water activity and temperature on the behaviour of A. parasiticus is similar to that described above for A. flavus. Pitt and Miscamble have reported a minimum for growth of about 0.83; and a minimum for aflatoxin production of about 0.87. There are only limited data on the effect of temperature on the growth of A. parasiticus and the production of the aflatoxins. It was reported that optimal growth and toxin production occur at approximately 30 and 28°C, respectively.

The term 'aflatoxins' was coined in the early 1960s when the death of thousands of turkeys ('Turkey X' disease), ducklings and other domestic animals was attributed to the presence of A. flavus toxins in groundnut meal imported from South America.

(Although the aflatoxins are the major toxins associated with this mycotoxicosis, another mycotoxin - cyclopiazonic acid - has been implicated in the aetiology of Turkey X disease.) The chronic effects of low dietary levels (parts per billion) of aflatoxin on livestock are also well documented and include decreased productivity and increased susceptibility to disease.

The aflatoxin-producing moulds occur widely, in temperate, sub-tropical and tropical climates, throughout the world; and the aflatoxins may be produced, both before and after harvest, on many foods and feeds especially oilseeds, edible nuts and cereals.

Although the aflatoxins are predominantly associated with commodities of sub-tropical and tropical origin, their occurrence has also been reported in temperate climates in acid-treated grains.

Aflatoxin B_1 is a human carcinogen and is one of the most potent hepatocarcinogens known. Human fatalities have also occurred from acute aflatoxin poisoning in India (in 1974), for example, when unseasonal rains and a scarcity of food prompted the consumption of heavily contaminated maize. If the immunosuppressive action of the aflatoxins in livestock is similarly manifested in humans, it is possible that the aflatoxins (and other mycotoxins) could play a significant role in the aetiology of human disease in some developing countries, where a high exposure to these toxins has been reported.

Lubulwa and Davis have studied economic losses attributable to the occurrence of aflatoxin only, in maize and groundnuts, in Southeast Asian countries (Thailand, Indonesia and the Philippines). They concluded that contaminated maize accounted for about 66 per cent of the total loss, whereas losses attributable to spoilage and deleterious effects on human and animal health were 24, 60 and 16 per cent of the total, respectively. However, the study considered losses associated with morbidity and premature death caused by cancer only. Consequently, it is likely that when the additional effects on human health caused by the immunotoxic effect of aflatoxin (and other mycotoxins) are included, the loss associated with aflatoxins will be significantly increased.

The Trichothecenes

Surprisingly little is known about the effects of water activity and temperature on the behaviour of the Fusarium moulds, including the production of mycotoxins.

In the case of F. graminearum, the temperature limits for growth have not been reported, although the optimal temperature has been estimated at 24 - 26 °C. The minimum water activity for growth is 0.9; the maximum limit is recorded as in excess of 0.99. No information is available on the effect of water activity and temperature on the production of deoxynivalenol, nivalenol and zearalenone.

The minimum water activity for the growth of F. sporotrichioides is 0.88, whereas the maximum limit is reported as >0.99. The minimum, optimal and maximum temperatures for growth are -2.0, 22.5 - 27.5 and 35°C, respectively. As with the other Fusariummoulds, there is no information on the conditions required for the production of T-2 toxin.

T-2 toxin and deoxynivalenol belong to a large group of structurally-related sesquiterpenes known as the 'trichothecenes'.

T-2 toxin is produced on cereals in many parts of the world and is particularly associated with prolonged wet weather at harvest. It is the probable cause of 'alimentary toxic aleukia' (ATA), a disease which affected thousands of people in Siberia during the Second World War, leading to the elimination of entire villages. The symptoms of ATA included fever, vomiting, acute inflammation of the alimentary tract and a variety of blood abnormalities. T-2 toxin is responsible for outbreaks of haemorrhagic disease in animals and is associated with the formation of oral lesions and neurotoxic effects in poultry. The most significant effect of T-2 toxin (and other trichothecenes) is the immunosuppressive activity which has been clearly demonstrated in experimental animals; and which is probably linked to the inhibitory effect of this toxin on the biosynthesis of macromolecules. There is limited evidence that T-2 toxin may be carcinogenic in experimental animals.

Deoxynivalenol (DON) is probably the most widely occurring Fusarium mycotoxin, contaminating a variety of cereals, especially maize and wheat, in both the developed and developing world. The outbreaks of emetic (and feed refusal) syndromes amongst livestock, caused by the presence of DON in feeds, has resulted in the trivial name, vomitoxin, being attributed to this mycotoxin.

The ingestion of DON has caused outbreaks of acute human mycotoxicoses in India, China and rural Japan. The Chinese outbreak, in 1984-85, was caused by mouldy maize and wheat; symptoms occurred within five to thirty minutes and included nausea, vomiting, abdominal pain, diarrhoea, dizziness and headache.

To date, nivalenol-producing isolates of F. graminearum have been observed, on rice and other cereals, only in Japan and have been associated with the occurrence of red mould disease ('Akakabi-byo'). Symptoms include anorexia, nausea, vomiting, headache, abdominal pain, diarrhoea and convulsions.

Zearalenone

Zearalenone is a widely distributed oestrogenic mycotoxin occurring mainly in maize, in low concentrations, in North America, Japan and Europe. However, high concentrations can occur in developing countries, especially when maize is grown under more temperate conditions in, for example, highland regions.

Zearalenone is co-produced with deoxynivalenol by F. graminearum and has been implicated, with DON, in outbreaks of acute human mycotoxicoses.

Exposure to zearalenone-contaminated maize has caused hyperoestrogenism in livestock, especially pigs, characterised by vulvar and mammary swelling and infertility. There is limited evidence in experimental animals for the carcinogenicity of zearalenone.

The Fumonisins

The fumonisins are a group of recently characterised mycotoxins produced by F. moniliforme, a mould which occurs worldwide and is frequently found in maize. Fumonisin B_1 has been reported in maize (and maize products) from a variety of agroclimatic regions including the USA, Canada, Uruguay, Brazil, South Africa, Austria, Italy and France. The toxins especially occur when maize is grown under warm, dry conditions.

The minimum water activity for the growth of F. moniliforme is 0.87; the maximum limit is recorded as >0.99. The minimum, optimal and maximum temperatures for growth are 2.5 - 5.0, 22.5 - 27.5 and 32 -37°C, respectively. There is no information on the conditions required for the production of fumonisin B_1.

Exposure to fumonisin B_1 (FB1) in maize causes leukoencephalomalacia (LEM) in horses and pulmonary oedema in pigs. LEM has been reported in many countries including the USA, Argentina, Brazil, Egypt, South Africa and China. FB1 is also toxic to the central nervous system, liver, pancreas, kidney and lung in a number of animal species.

The presence of the fumonisins in maize has been linked with the occurrence of human oesophageal cancer in the Transkei, southern Africa and China. The relationship between exposure to F. moniliforme, in home-grown maize, and the incidence of oesophageal cancer has been studied in the Transkei during the ten-year period 1976-86. The percentage of kernels infected by F. moniliforme was significantly higher in the high-risk cancer area during the entire period; and FB1 and FB2 occurred at significantly higher levels in mouldy maize obtained from high-risk areas in 1986.

Previously, an evaluation by the International Agency for Research on Cancer had concluded that there is sufficient evidence in experimental animals for the carcinogenicity of cultures of F. moniliforme that contain significant amounts of the fumonisins; whereas there is limited evidence, in experimental animals, for the carcinogenicity of fumonisin B_1. However, the results of a recently completed study of the toxicology and carcinogenesis of fumonisin B_1 has been reported by the National Toxicology Program of the US Department of Health and Human Services.

Although the report is still in draft form, it concludes that there is clear evidence of carcinogenic activity of fumonisin B_1 in male F344/N rats based on the increased incidences of renal tubule neoplasms; and that there is also clear evidence of carcinogenic activity of fumonisin B_1 in female $B6C3F_1$ mice based on the increased incidences of hepatocellular neoplasms. There is no evidence of carcinogenic activity of fumonisin B_1 in female rats or male mice.

Ochratoxin A

A. ochraceus grows more slowly than both A. flavus and A. parasiticus, but can grow at a water activity as low as 0.79. Growth has also been reported within the temperature range 8 - 37 °C, with an optimum variously reported as 25 - 31°C. Ochratoxin A is produced within the temperature range 15 - 37 °C, with an optimal production at 25 - 28 °C.

P. verrucosum grows within the temperature range 0 - 31°C and at a minimum water activity of 0.80. Ochratoxin A is produced over the whole temperature range. Significant quantities of toxin can be produced at a temperature as low as 4°C, and at a water activity as low as 0.86.

Exposure to ochratoxin A (OA) appears to occur mainly in wheat and barley growing areas in temperate zones of the northern hemisphere. The levels of OA reported in these commodities ranges from trace amounts to 6000 μg/kg, in Canadian wheat. In the UK, reported levels have varied from <25 to 5,000 and from <25 to 2,700 μg/kg in barley and wheat respectively. It also occurs in maize, rice, peas, beans, cowpeas, vine fruits and their products, coffee, spices, nuts and figs.

The ability of OA to transfer from animal feeds to animal products has been demonstrated by the occurrence of this toxin in retail pork products, and the blood of swine, in Europe.

Although cereal grains are considered to be the main human dietary source of OA, it has been suggested that pork products may also be a significant source of this toxin. Ochratoxin A has been found in blood (and milk) from individuals in a variety of European countries, including France, Italy, Germany, Denmark, Sweden, Poland, Yugoslavia and Bulgaria. One of the highest reported levels is 100 ng/ml OA in blood from Yugoslavia ; whereas 6.6 ng/ml OA in milk has been recorded in Italy.

Existing or proposed regulations for OA are available in at least eleven countries, the permitted levels ranging from 1 to 50 μg/kg in foods and from 100 to 1000 μg/kg in feeds. In Denmark, the acceptability of pork products from a specific carcass is determined by analysing the OA content of the kidney. The pork meat and certain organs can be consumed as food if the OA content of the kidney is no more than 25 and 10 μg/kg respectively.

A provisional tolerable weekly intake of 100 ng/kg bw, of OA, approximating to 14 ng/kg body weight per day, has been recommended by a WHO/FAO Joint Expert Committee on Food Additives, JECFA.

Ochratoxin A has been linked with the human disease Balkan endemic nephropathy, a fatal, chronic renal disease occurring in limited areas of Bulgaria, the former Yugoslavia and Romania. OA causes renal toxicity, nephropathy and immunosuppression in several animal species and it is carcinogenic in experimental animals.

Patulin

Patulin is an antibiotic produced by a number of moulds. It occurs in rotten apples contaminated by Penicillium expansumand, consequently, may occur in apple juice and other apple-based products.

Experimental studies have demonstrated that patulin is a neurotoxin and that it produces marked pathological changes in the viscera. Although patulin has been reported as inducing local sarcomas, no mutagenic activity has been discernible in most short-term tests.

JECFA has established a provisional maximum tolerable daily intake of 400 ng/kg bw for patulin.

Co-occurrence of Mycotoxins

The complex ecology of mould growth and mycotoxin production can produce mixtures of mycotoxins in foods and feeds, especially in cereals.

The co-occurrence of mycotoxins can affect both the level of mycotoxin production and the toxicity of the contaminated material. The production of the aflatoxins in stored grains, for example, may be enhanced by the presence of trichothecenes, whereas the toxicology of naturally occurring combinations of trichothecene mycotoxins is reportedly determined by synergistic interactions, in experimental animals. For example, in a study with swine, the affect of deoxynivalenol on weight gain and feed conversion was synergized by T-2 toxin. Interactions involving non-toxic fungal metabolites have also been reported, including the potent synergism of non-toxic F. graminearum metabolites (culmorin, dihydroxycalonectrin and sambucinol) with deoxynivalenol. To date, too little is known about this particularly important area of mycotoxicology.

Mycotoxins of Regional Importance

There are a number of mycotoxicoses which are not widely occurring, but which are of importance to the exposed populations in the affected regions. Mycotoxicoses which fall into this category include those associated with moulds occurring in both growing and stored forage crops. The moulds and mycotoxins include those which have been associated with a variety of livestock diseases including ergotism, paspalum staggers, ryegrass staggers, facial eczema, fescue foot, lupinosis, slobber syndrome and stachybotryotoxicosis.

Table 2. Moulds and mycotoxins of regional importance

Mould species	*Mycotoxins produced*	*Mycotoxicosis*
Claviceps purpurea	Ergotamine alkaloids	Ergotism
Claviceps fusiformis	Clavine alkaloids	Ergotism
Claviceps paspali	Paspalinine	Paspalum staggers
Acremonium loliae	Lolitrem	Ryegrass staggers
Balansia spp	Alkaloids	Fescuefoot
Pithomyces chartarum	Sporidesmin	Facial eczema
Phomopsis leptostromiformis	Phomopsin	Lupinosis
Rhizoctonia leguminicola	Slaframine	Slobber syndrome
Stachybotrys atra	Satratoxins	Stachybotryotoxicosis
Diplodia maydis	Diplodiatoxin	Diplodiosis

Most farm animals consume pasture crops, either by grazing on the living pasture or by consuming the crops as hay or silage. The crops can be colonized by moulds throughout this period, the development of the moulds and the production of fungi being dependent on the prevailing ecosystem. Growing crops present a variety of micro-environments. For example, the uppermost leaves of a plant will be subjected to extreme fluctuations of temperature and relative humidity, whereas those leaves towards the base of the plant will present a more shaded, moderate, humid environment; the surface texture of the leaf will also effect the micro-environment.

The Socio-Economic System

The socio-economic system describes those social (eg cultural, political) and economic (macro- and micro) factors which will exert an important influence on events within the mycotoxicology system; and which should be most thoroughly addressed when any attempt is made to control the production of moulds and mycotoxins. In some instances, given the complexity and unpredictability of human behaviour, it can be very difficult to intervene successfully within the socio-economic system. However, technical interventions which are designed to alleviate spoilage will only be successfully implemented if they can be accommodated and exploited within the existing socio-economic system. Whenever efforts are made to improve the quality of foods and feeds, it should be clearly established that there is a definite need for a better quality product, and that the community is prepared to bear any associated increase in the cost of the improved commodity.

The Control System

The successful management of interacting commodity systems ('commodity management') requires the co-ordinated inputs of aninterdisciplinary team', where the potential advantages arising from the dynamics of the team are realised by fully exploiting theinteractions between the skills, disciplines and backgrounds of the individual team members. The team will have the skills required to enable it to operate across commodity systems, identifying those factors which are compromising the quality of the products, and introducing appropriate interventions.

The Control System illustrates a selection of preventative and curative interventions (measures) which may be utilised for the control of mycotoxins, once the nature of the contamination process has been properly evaluated.

Those factors which are compromising the quality of the products of the commodity system, and leading to the production of moulds and mycotoxins, may be evaluated by the implementation of: carefully designed surveillance studies; recently developed biomonitoring methods, to measure the exposure to mycotoxins of individuals; and socio-economic studies, which address a variety of social, marketing and financial issues. The occurrence of moulds and mycotoxins can be alleviated by the application of a variety of preventative measures both before and after harvest including, for example, appropriate pest and disease control measures and good harvesting, drying and storage practices. Once mycotoxin contamination has occurred, it can be alleviated by a variety of predominantly post-harvest measures including processing, detoxification, and segregation. A structured, systematic approach to the control of mycotoxins is required, focusing upon the need for preventative control measures, and recognising the intimate interactions that occur throughout commodity systems and related systems.

Integration of Systems

Hazard Analysis and Critical Control Point (HACCP) is a food safety management system that is based upon the systematic identification and assessment of hazards in foods, and the definition of means to control them. It is an important component of an integrated approach to food safety.

Economic Implications of Mycotoxins

International trade in agricultural commodities such as wheat, rice, barley, corn, sorghum, soybeans, groundnuts and oilseeds amounts to hundreds of millions of tonnes each year. Many of these commodities run a high risk of mycotoxin contamination. Regulations on mycotoxins have been set and are strictly enforced by most importing countries, thus affecting international trade. For some developing countries, where agricultural commodities account for as much as 50 percent of the total national exports, the economic importance of mycotoxins is considerable.

There is a notable length of time between the purchase of the agricultural commodity at the village market of the exporting country and its arrival at the distribution centre of the importing country. The transport chain for agricultural commodities from village to port, from the port of the exporting country to that of the importing country as well as from the port to the distribution centre can be long. Furthermore, storage conditions at the farm level as well as during transport under adverse weather conditions may not always be satisfactory. For these reasons, there is considerable opportunity for mould and mycotoxin contamination of agricultural commodities to take place throughout the food system - from production to distribution and transport - and this may lead to economic losses.

The exact figures for world economic losses resulting from mycotoxin - contamination may never be available. Apart from the obvious losses of food and feed, there are losses caused by lower productivity; losses of valuable foreign exchange earnings; costs incurred by inspection, sampling and analysis before and after shipments; losses attributable to compensation paid in case of claims; farmer subsidies to cover production losses; research, training and extension programme costs; costs of detoxification; etc. When combined, these costs may be staggering.

Prevention and Control

Several preventive measures to minimize mycotoxin contamination in agricultural commodities have been attempted. These can be divided into three broad categories:

— plant breeding;

— good agronomic practices; and

— detoxification.

The problem of ergot contamination of cereals and millets has been successfully minimized in the past by cultivating varieties of rye, wheat and pearl millet that are resistant to the disease. The periodic re-emergence of the problem can be attributed to the release of varieties which are not resistant to ergot. However, there has been little success in providing resistant varieties of corn and peanut to minimize the problem of aflatoxins. Other agronomic approaches such as avoiding water stress, minimizing insect infestation and reducing inoculum potential have been suggested and are effective when the farmers can implement such practices. Following good agricultural practices during both pre-harvest and post-harvest conditions would, minimize the problem of contamination by mycotoxins such as aflatoxins, ochratoxin and trichothecene mycotoxins. These include appropriate drying techniques, maintaining proper storage facilities and taking care not to expose the grains or oilseeds to moisture during transport and marketing. The method of segregating contaminated, mouldy, shrivelled or insect-infested seeds from sound kernels has been particularly useful in minimizing aflatoxin contamination in peanuts. Detoxification of aflatoxins in foods and animal feeds has been attempted in the past. For instance, Senegal has been operating commercial facilities to detoxify peanut cake contaminated with aflatoxins by the ammonia process. Any detoxification procedure must be tested for safety and efficacy and invariably results in increased handling and costs. In addition, the detoxified product has been considered suitable only for animal feed purposes and not for human consumption.

Several countries have introduced legislation concerning mycotoxins. Most of this legislation pertains to aflatoxins, ergot alkaloids, deoxynivalenol and ochratoxins. Although various legislative measures have yet to be harmonized among countries, the Codex Alimentarius Commission is making efforts to establish international guideline levels for mycotoxins, and aflatoxins in particular.

Mycotoxin control measures have been implemented for agricultural commodities entering international trade or located in countries with centralized or large-scale buying and distribution systems. However, in developing countries, where local food consumption or subsistence agriculture is practised by as much as 70 percent of the population, such measures would be difficult to implement. Because of the stringent mycotoxin control measures being maintained by those countries importing food grains and oilseeds, and because of the need for exporting countries to earn foreign exchange, the best of the commodities are often sold abroad while the substandard or contaminated commodities are retained for domestic use. Such a situation exacerbates the risk to human and animal health.

Continuous surveillance of high-risk agricultural commodities for contamination by selected mycotoxins and the monitoring of human population groups for diseases attributable to mycotoxins have to be carried out throughout the world to ensure a supply of safe food which is free from naturally occurring contaminants. The financial and human investments in this endeavour would be returned in terms of better human and animal health as well as reduced economic losses.

Hazard Analysis and Critical Control Point (HACCP)

HACCP was originally developed as a microbiological safety system in the early days of the US manned space programme in order to guarantee the safety of astronauts' food. Up until that time most food safety systems were based on end product testing and could not fully assure safe products as 100% testing was impossible. A pro-active, process-focused system was needed and the HACCP concept was born.

The original system was designed by the Pilsbury Company working alongside NASA and the US army laboratories at Natick. It was based on the engineering system Failure, Mode and Effect Analysis (FMEA) which looked at what could potentially go wrong at each stage in the operation along with possible causes and the likely effect, before applying effective control mechanisms.

HACCP is a system that identifies, evaluates and controls hazards which are significant for food safety. It is a structured, systematic approach for the control of food safety throughout the commodity system, from the plough to the plate. It requires a good understanding of the relationship between cause and effect in order to be more pro-active and it is a key element in Total Quality Management (TQM). HACCP builds on the foundations of well established quality management systems such as Good Manufacturing Practice (GMP), Good Hygienic Practice (GHP), Good Agricultural Practice (GAP), and Good Storage Practice (GSP). The HACCP concept has been successfully applied in the control of quality as well as safety in low-acid canned foods in the USA, and many food companies in Europe and the USA have adopted the approach. Increasingly, regulatory bodies have recognised the usefulness of this tool and its 'principles' have has been incorporated into legislative requirements by both the EU (in the General Hygiene regulations for managing food safety (93/43/EEC)), and the United States Federal Department of Agriculture (CPR - 123). The National Advisory Committee on Microbiological Criteria for Foods (NACMCF) provided guidelines on HACCP including generic plans and decision trees in 1992, and the Codex Alimentarius Commission adopted the HACCP system at its twentieth session in 1993. HACCP systems can be incorporated into other quality assurance systems such as the ISO 9000 series.

Although conceived as a food safety system for both the agricultural and processing systems, it is in the latter that HACCP has found most application hitherto. This is primarily because it is much easier to apply a HACCP system in a factory where there is a single management or 'owner', and where it is possible to completely prevent a food safety hazard, or eliminate, or reduce it to an acceptable level. In the commodity system there are often many disparate 'owners' of the commodity as it passes from the farm to the consumer, and complete control may be unobtainable. This Manual aims to address this subject, basing the approach as closely as possible on the Codex Code of General Principles on Food Hygiene, which emphasises the importance of GMP/GAP/GHP as sound foundations to incorporate the HACCP approach and develop a user friendly Food Safety Management System.

Pre-requisite Programmes

Pre-requisite programmes such as GAP, GMP and GHP must be working effectively within a commodity system before HACCP is applied. If these pre-requisite programmes are not functioning effectively then the introduction of HACCP will be complicated, resulting in a cumbersome, over-documented system.

Good Agricultural Practices

Primary Production

Primary food production should be managed to ensure that food is safe and wholesome for the consumer. Production will start on the farm, in the sea or lake or even within a forest. It is essential that certain ground rules are followed. Land used for crop or horticulture production should be fit for purpose and should not have previously been contaminated with heavy metals, industrial chemicals or environmental waste. Such hazards will be transferred into the food chain rendering the commodity unfit for human consumption. Farmers should control production so that contamination of the crop, proliferation of pests, and diseases of animals and plants, do not compromise food safety. Good Agricultural Practices (GAP), including Good Hygienic Practices (GHP) where appropriate, should be adopted to make sure that the harvested commodity will not present a food hazard to the consumer. Good Storage Practices (GSP) should be followed when the commodity is stored on the farm. As well as being covered in Food Hygiene Basic Texts (CODEX) there are also four ISO procedures that cover the storage of cereal s and pulses (ISO 6322 series). GSP should also be followed for storage throughout the commodity system.

Good Manufacturing Practices

Establishment Design and Facilities

The structure and location of a processing plant needs to be considered in relation to the nature of operations and risks associated with them.

- Food premises should be designed to minimise possibilities of contamination of commodity or product.
- Design and layout should permit maintenance, cleaning and disinfection of the site to minimise airborne contamination.
- All surfaces that come into contact with food should be non toxic, as well as being easy to maintain and clean in order to prevent any additional contamination.
- Suitable facilities should exist for temperature and humidity control, when required.
- Effective measures should exist to prevent access by pests

Control of Operation

Effective control measures should be in place to reduce the risk of contamination of the commodity or food supply such that it is safe and fit for purpose:

- Adequate time, temperature or humidity controls
- Food grade packaging
- Potable water supplies
- Maintenance of equipment

Maintenance and Sanitation

Procedures and work instructions should exist to demonstrate an adequate level of maintenance of an establishment as well as efficient practices for cleaning, waste management, and pest control. Overall, these operations will support the ongoing control of potential food hazards that may contaminate food.

Personnel Hygiene

Measures need to be in place to ensure that food handlers do not contaminate food. This objective can be attained by maintaining an appropriate level of personal cleanliness and following guidelines for personal hygiene.

Transportation

The method of transportation should be such that measures are taken to prevent any contamination or deterioration of the commodity. Commodities or product that need to be transported in certain environments should be appropriately controlled, e.g. chilled, frozen, or stored under specific humidity levels.

Containers and conveyors used for transporting food need to be maintained in good condition and be easy to clean.

Containers used for bulk transfer should be designated and marked specifically for food use only.

Training

All food handlers should be trained in personal hygiene, as well as in the specific operation with which they are working, to a level commensurate with their duties. Food handlers should also be supervised by trained supervisors.

An ongoing training programme for food handlers is paramount to the success of a Food Safety Management System

Product Information and Consumer Awareness

The end product should be accompanied by adequate information to ensure that personnel at the next stage in the food chain will handle, store, process, prepare and display the product safely. Since the consumer may be responsible for performing the ulitimate control measure, the cooking of raw meat or fish, they should have all the relevant information required to carry out this step effectively.

All batches of food should be easily identified, by a batch or lot number, to allow traceability of the commodity if required.

Basic Principles of HACCP

There are seven discrete activities that are necessary to establish, implement and maintain a HACCP plan, and these are referred to as the 'seven principles' in the Codex Guideline.

The seven principles are:

1. *Principle 1.Conduct a hazard analysis.* Identify hazards and assess the risks associated with them at each step in the commodity system. Describe possible control measures.

2. *Principle 2. Determine the Critical Control Points (CCPs).* A critical control point is a step at which control can be applied and is essential to prevent or eliminate a food safety hazard, or reduce it to an acceptable level.
3. *Principle 3. Establish critical limits.* Each control measure associated with a CCP must have an associated critical limit which separates the acceptable from the unacceptable control parameter.
4. *Principle 4. Establish a monitoring system.* Monitoring is the scheduled measurement or observation at a CCP to assess whether the step is under control, i.e. within the critical limit(s) specified in Principle 3.
5. *Principle 5. Establish a procedure for corrective action, when monitoring at a CCP indicates a deviation from an established critical limit.*
6. *Principle 6. Establish procedures for verification to confirm the effectiveness of the HACCP plan.* Such procedures include auditing of the HACCP plan to review deviations and product dispositions, and random sampling and checking to validate the whole plan.
7. *Principle 7. Establish documentation concerning all procedures and records appropriate to these principles and their application*

Developing a HACCP Plan

There are twelve tasks required to develop a HACCP plan and these are designed to ensure that the seven principles are applied correctly. Principle 1, which is to conduct a hazard analysis, requires that the first five tasks have all been addressed in a logical and honest manner so that all real hazards associated with the commodity have been identified.

Task 1 - Establish a HACCP team

To fully understand the commodity system and be able to identify all likely hazards and CCPs, it is important that the HACCP team is made up of people from a wide range of disciplines. The team should include:

- A team leader to convene the group and to direct the work of the team ensuring that the concept is properly applied. This person must be familiar with the technique, be a good listener and allow all participants to contribute.

- A specialist with a detailed knowledge of the commodity system is required. This specialist will have a major role in the production of the commodity flow diagrams.
- Several specialists, each with an understanding of particular hazards and associated risks.
- People, such as packaging specialists, raw material buyers, distribution staff or production staff, farmers, brokers, who are involved with the process, and have working knowledge of it, may be brought into the team temporarily in order to provide relevant expertise.
- The team's progress and results of the analysis should be recorded by a technical secretary.

If any changes are made to composition or operational procedures, it will be necessary to re-assess the HACCP plan in the light of the changes.

The first activity of the HACCP team is to identify the scope of the study. For example, will the whole commodity system be covered, or only selected components? This will make the task more manageable and specialists can be added to the team as and when they are required.

Task 2 - Describe the product

To start a hazard analysis, a full description of the product, including customer specification, should be prepared. This should include information relevant to safety, e.g. mycotoxin regulation/target level, composition, physical/chemical properties of the raw materials and the final product, the amount of water available for microbial growth (a_w), the amount of acid or alkali in the product (pH). Also information regarding how the product is to be packaged, stored and transported should also be considered together with facts regarding its' shelf life and recommended storage temperatures. Where appropriate, labelling information and an example of the label should be included. This information will help the HACCP team to identify 'real' hazards associated with the process.

Task 3 - Identify the product's intended use

How the product is intended to be used is an important consideration. Information on whether the product will be consumed directly, or be cooked, or be further processed, will all have a bearing on the hazard analysis. The nature of the target group for the product may also be relevant, particularly if it includes susceptible groups such as infants, the elderly, and the malnourished. The likelihood of misuse of a product should also be considered, such as the use of pet food as a human food, either by accident or design.

Task 4 - Draw up the commodity flow diagram

The first function of the team is to draw up a detailed commodity flow diagram (CFD) of the commodity system, or that part of it which is relevant. The expertise of the commodity specialist is important at this stage. Commodity systems will differ in detail in different parts of the world, and even within one country there may be a number of variants. Secondary processing will need to be detailed for each factory, using generic flows only as a guide.

Task 5 - On site confirmation of flow diagram

Upon completion of the CFD, members of the team should visit the commodity system (e.g. farm, store or manufacturing area) to compare the information present on the CFD with what actually happens in practice. This is known as "walking the line", a step by step practice to check that all information regarding materials, practices, controls etc. have been taken into consideration by the team during the preparation of the CFD. Information such as time of harvest, drying procedures, storage conditions, the marketing chain, socio-economic factors, grading systems and any incentive for improved quality or safety, and processing systems, should be collected and included in the CFD as appropriate. The site for which the HACCP plan is being designed should be visited as many times as possible to ensure that all relevant information has been collected.

Task 6 - Identify and analyse hazard(s)

Effective hazard identification and hazard analysis are the keys to a successful HACCP Plan. All real or potential hazards that may occur in each ingredient and at each stage of the commodity system should be considered. Food safety hazards for HACCP programmes have been classified into three types of hazards:

- Biological: typically foodborne bacterial pathogens such as Salmonella, Listeria and E. coli, also viruses, algae, parasites and fungi.
- Chemical: There are three principle types of chemical toxins found in foods: naturally occurring chemicals, e.g. cyanides in some root crops, and allergenic compounds in peanuts; toxins produced by micro-

organisms, e.g. mycotoxins, and algal toxins; and chemicals added to the commodity by man to control an identified problem, e.g fungicides or insecticides.

- Physical: contaminants such as broken glass, metal fragments, insects or stones.

The probability that a hazard will occur is called a risk. The risk may take a value from zero to one depending on the degree of certainty that the hazard will be absent or that it will be present. After hazard identification, a hazard analysis must be conducted to understand the relative health risk to man or animal posed by the hazard. It is a way of organizing and analyzing the available scientific information on the nature and size of the health risk associated with the hazard. The risk may have to be assessed subjectively and simply classified as low, medium, or high.

Once a food safety hazard has been identified, then appropriate control measures should be considered. These are any action or activity that can be used to control the identified hazard, such that it is prevented, eliminated, or reduced to an acceptable level. The control measure may also include training of personnel for a particular operation, covered by GAP, GMP, and GHP.

Task 7 - Determine the critical control points (ccps)

Each step in the commodity flow diagram, within the scope of the HACCP study, should be taken in turn and the relevance of each identified hazard should be considered. It is also important to remember the stated scope of the HACCP analysis at this stage. The team must determine whether the hazard can occur at this step, and if so whether control measures exist. If the hazard can be controlled adequately, and is not best controlled at another step, and is essential for food safety, then this step is a CCP for the specified hazard.

If a step is identified where a food safety hazard exists, but no adequate control measures can be put in place either at this step or subsequently, then the product is unsafe for human consumption. Production should cease until control measures are available and a CCP can be introduced.

Task 8 - Establish critical limits for each ccp

Critical limits must be specified and validated for each CCP. Criteria often used include measurements of temperature, time, moisture level, pH, water activity, and sensory parameters such as visual appearance. In the case of mycotoxins for example, they may include the moisture content or the temperature of the commodity. All critical limits, and the associated permissible tolerances, must be documented in the HACCP Plan Worksheet, and included as specifications in operating procedures and work instructions.

Task 9 - Establish a monitoring procedure

Monitoring is the mechanism for confirming that critical limits at each CCP are being met. The method chosen for monitoring must be sensitive and produce a rapid result so that trained operatives are able to detect any loss of control of the step. This is imperative so that corrective action can be taken as quickly as possible so that loss of product will be avoided or minimised.

Monitoring can be carried out by observation or by measurement, on samples taken in accordance with a statistically based sampling plan. Monitoring by visual observation is basic but gives rapid results, and can therefore be acted upon quickly. The most common measurements taken are time, temperature and moisture content.

Task 10 - Establish corrective action

If monitoring indicates that critical limits are not being met, thus demonstrating that the process is out of control, corrective action must be taken immediately. The corrective action should take into account the worst case scenario, but must also be based on the assessment of hazards, risk and severity, and on the final use of the product. Operatives responsible for monitoring CCPs should be familiar with and have received comprehensive training in how to effect a corrective action.

Corrective actions must ensure that the CCP has been brought back under control. They must also include appropriate disposition of any affected commodity or product. Whenever possible an alarm system should be introduced which will activate when monitoring indicates that the critical limit is being approached. Corrective action can then be applied to pre-empt a deviation and prevent the need for any product disposition.

Task 11 - Verify the HACCP plan

Once the HACCP plan has been drawn up, and all of the CCPs have been validated, then the complete plan must be verified. Once the HACCP plan is in routine operation, it must be verified and reviewed at regular intervals.

This should be a task of the person charged with the responsibility for that particular component of the commodity system The appropriateness of CCPs and control measures can thus be determined, and the extent and effectiveness of monitoring can be verified. Microbiological and/or alternative chemical tests can be used to confirm that the plan is in control and the product is meeting customer specifications. A formal internal auditing plan of the system will also demonstrate an ongoing commitment to keep the HACCP plan up to date, as well as representing an essential verification activity.

Ways in which the system can be verified include:

- collecting samples for analysis by a method different from the monitoring procedure
- asking questions of staff, especially CCP monitors
- observing operations at CCPs
- formal audit by independent person

It is important to remember that the HACCP system is set up for a particular formulation of product handled and processed in a given way.

Task 12 - Keep record

Record keeping is an essential part of the HACCP process. It demonstrates that the correct procedures have been followed from the start to the end of the process, offering product traceability. It provides a record of compliance with the critical limits set, and can be used to identify problem areas. Furthermore, the documentation can be used by a company as evidence of 'Due Diligence Defence' as required, for instance, by the Food Safety Act 1990 (HMSO), in the UK.

Records that should be kept include: all processes and procedures linked to GMP, GHP, CCP monitoring, deviations, and corrective actions.

Documents should also include those that recorded the original HACCP study, e.g. hazard identification and selection of critical limits, but the bulk of the documentation will be records concerned with the monitoring of CCPs and corrective actions taken. Record keeping can be carried out in a number of ways, ranging from simple check-lists, to records and control charts. Manual and computer records are equally acceptable, but a documentation method should be designed that is appropriate for the size and nature of the enterprise.

References

Berry, L. 1988. The pathology of mycotoxins. J. *Pathology,* 154: 301-311.

Bhat, R.V., Sashidhar, R.B., Ramakrishna.Y. & Munshi, K.L. 1989. Outbreak of trichothecene mycotoxicosis associated with consumption of mould damaged wheat products in Kashmir Valley, India. *Lancet,* I: 35-37.

Jelinek, C.F., Pohland, A.E. & Wood, G.E. 1989. Worldwide occurrence of mycotoxins in foods and feeds (an update. *J. Assoc. Off. Anal. Chem.,* 72: 223-230.

Palmgren, M.S. & Hayes, A.W. 1987. Aflatoxins in food. In P. Krogh, ed., *Mycotoxins in food,* p. 65-95, London, Academic Press.

Rodricks, J.V. & Stoloff, L. 1977. Aflatoxin residues from contaminated feed in edible tissues of food-producing animals. *In* J.V. Rodricks, C.W. Hesseltine & M.A. Mehlman, eds., *Mycotoxins and animal health,* p, 67-79, Park Forest Hills, Illinois, Pathotox Publishers Inc.

6

Probiotics

The term probiotic is a relatively new word meaning "for life" and it is currently used to name bacteria associated with beneficial effects for humans and animals. The original observation of the positive role played by some selected bacteria is attributed to Eli Metchnikoff, the Russian born Nobel Prize winner working at the Pasteur Institute at the beginning of the last century, who suggested that "The dependence of the intestinal microbes on the food makes it possible to adopt measures to modify the flora in our bodies and to replace the harmful microbes by useful microbes".

At this time Henry Tissier, a French paediatrician, observed that children with diarrhea had in their stools a low number of bacteria characterized by a peculiar, Y-shaped morphology. These "bifid" bacteria were, on the contrary, abundant in healthy children. He suggested that these bacteria could be administered to patients with diarrhea to help restore a healthy gut flora.

The works of Metchnikoff and Tissier were the first to make scientific suggestions concerning the probiotic use of bacteria, even if the word "probiotic" was not coined until 1960, to name substances produced by microorganisms which promoted the growth of other microorganisms. Fuller, in order to point out the microbial nature of probiotics, redefined the word as "A live microbial feed supplement which beneficially affects the host animal by improving its intestinal balance". A similar definition was proposed by Havenaar and Huis in 't Veld, "a viable mono or mixed culture of bacteria which, when applied to animal or man, beneficially affects the host by improving the properties of the indigenous flora". A more recent, but probably not the last definition is "live microorganisms, which when consumed in adequate amounts, confer a health effect on the host".

It is clear that these definitions have:

1) restricted the use of the word probiotic to products which contain live microorganisms;
2) pointed out the need for providing an adequate dose of probiotic bacteria in order to exert the desirable effects.

The observations of Metchnikoff and Tissier were so appealing that commercial exploitation immediately followed their scientific works. Unfortunately, results were not always positive and most of these observations were anecdotal. The probiotic concept was therefore regarded as scientifically unproven and it received minor interest for decades, with some research involving animal feeding, in order to find healthy substitutes for growth promoting agents. In the last 20 years however, research in the probiotic area has progressed considerably and significant advances have been made in the selection and characterization of specific probiotic cultures and substantiation of health claims relating to their consumption.

Members of the genera Lactobacillus and Bifidobacterium are mainly used, but not exclusively, as probiotic microorganisms and a growing number of probiotic foods are available to the consumer. Some ecological considerations on the gut flora are necessary to understand the relevance, for human health, of the probiotic food concept.

Bacteria are normal inhabitants of humans (as well as the bodies of upper animals and insects) including the gastrointestinal tract, where more than 400 bacterial species are found : half of the wet weight of colonic material is due to bacterial cells whose numbers exceed by 10-fold the number of tissue cells forming the human body. Normally the stomach contains few bacteria (10^3 colony forming units per ml of gastric juice) whereas the bacterial concentration increases throughout the gut resulting in a final concentration in the colon of 10^{12} bacteria/g. Bacterial colonization of the gut begins at birth, as new-borns are maintained in a sterile status until the delivery begins, and continues throughout life, with notable age-specific changes. Bacteria, forming the so-called resident intestinal microflora, do not normally have any acute adverse effects and some of them have been shown to be necessary for maintaining the well being of their host.

As an example of the beneficial role of intestinal microflora, it is possible to cite what has been referred to as "colonization resistance" or "barrier effect" meaning the mechanism used by bacteria already present in the gut to maintain their presence in this environment and to avoid colonization of the same intestinal sites by freshly ingested microorganisms, including pathogens. Therefore, it could be assumed that dietary manipulation of gut microflora, in order to increase the relative numbers of "beneficial bacteria" could contribute to the well being of the host. This was also the original assumption of Metchnikoff who however, cautioned that:

> Systematic investigations should be made on the relation of gut microbes to precocious old age, and on the influence of diets which prevent intestinal putrefaction in prolonging life and maintaining the forces of the body.

This prudent statement can still be regarded today as an invitation to scientists to investigate the probiotic bacteria in more depth and with care.

For use in foods, probiotic microorganisms should not only be capable of surviving passage through the digestive tract but also have the capability to proliferate in the gut. This means they must be resistant to gastric juices and be able to grow in the presence of bile under conditions in the intestines, or be consumed in a food vehicle that allows them to survive passage through the stomach and exposure to bile. They are Gram positive bacteria and are included primarily in two genera, Lactobacillus and Bifidobacterium.

Selection of Probiotic Strains for Human Use

Probiotics must be able to exert their benefits on the host through growth and/or activity in the human body. However, it is the specificity of the action, not the source of the microorganism that is important. Indeed, it is very difficult to confirm the source of a microorganism. Infants are born with none of these bacteria in the intestine, and the origin of the intestinal microflora has not been fully elucidated. It is the ability to remain viable at the target site and to be effective that should be verified for each potentially probiotic strain. There is a need for refinement of in vitro tests to predict the ability of probiotics to function in humans. The currently available tests are not adequate to predict the functionality of probiotic microorganisms in the intestine.

Identification of Individual Strains

Classification is the arranging of organisms into taxonomic groups (taxa) on the basis of similarities or relationships. Nomenclature is the assignment of names to the taxonomic groups according to rules. Identification is the process of determining that a new isolate belongs to one of the established, named taxa. Probiotics be named according to the International Code of Nomenclature to ensure understanding on an international basis.

Since probiotic properties are strain related, it is suggested that strain identification (genetic typing) be performed, with methodology such as pulse field gel electrophoresis (PFGE). It is recommended that phenotypic tests be done first, followed by genetic identification, using such methods as DNA/DNA hybridization, 16S RNA sequencing or other internationally recognized methods. For the latter, the RDP (ribosomal data base project) should be used to confirm identity.

Measuring the Health Benefits of Probiotics

A number of health effects are associated with usage of probiotics. There are differing degrees of evidence supporting the verification of such effects and there are reports showing no clinical effects of certain probiotic strains in specific situations.

The use of probiotic microorganisms to confer health benefits on the host must indicate the dosage regimens and duration of use as recommended by the manufacturer of each individual strain or product based upon scientific evidence, and as approved in the country of sale. While this practice is not currently in place, each product should indicate the minimum daily amount required for it to confer specific health benefit(s). Such evidence should, where possible result from in vitro, animal (where appropriate) and human studies. Examples have been cited below to illustrate studies on specific strains and clinical outcomes. In doing so, the emphasis should not be on one particular strain being termed as superior to another, rather that the benefit conferred and the methods used to obtain and measure said benefits are of most importance.

Disorders Associated with the Gastrointestinal Tract

Prevention of diarrhea caused by certain pathogenic bacteria and viruses

Infectious diarrhea is a major world health problem, responsible for several million deaths each year. While the majority of deaths occur amongst children in developing countries, it is estimated that up to 30% of the population even in developed countries are affected by foodborne diarrhea each year. Probiotics can potentially provide an important means to reduce these problems. It should be noted that some of the studies referenced below utilize probiotics administered in a non-food form.

The strongest evidence of a beneficial effect of defined strains of probiotics has been established using Lactobacillus rhamnosus GG and Bifidobacterium lactis BB-12 for prevention and treatment of acute diarrhea mainly caused by rotaviruses in children.

In addition to rotavirus infections, many bacterial species cause death and morbidity in humans. There is good in vitro evidence that certain probiotic strains can inhibit the growth and adhesion of a range of enteropathogens, and animal studies have indicated beneficial effects against pathogens such as Salmonella. There is evidence from studies on travelers' diarrhea, where some of the causative pathogens have been presumed to be bacterial in nature, that benefits can accrue with probiotic administration.

It is important to note that probiotic therapy of acute diarrhea should be combined with rehydration if available. Current WHO recommendations state that clinical management of acute diarrhea should include replacement of fluid and electrolytes losses along with nutritional support. Oral rehydration salts (ORS) have been widely used in such disease management, and it is within this context that the combination therapy with probiotics is hereby advocated. Effects such as probiotic restoration of the non-pathogen dominated intestinal microflora secondary to infection, maintaining mucosal integrity and improving electrolyte balance could have a significant impact on programmes of treatment and prevention of acute diarrhea in developing countries.

A major problem associated with antibiotic treatment is the appearance of diarrhea, often caused by Clostridium difficile. This organism is not uncommon in a healthy intestinal tract, but the disruption of the indigenous microflora by antibiotics leads to an abnormal elevation of their numbers, and subsequent symptoms related to toxin production. The rationale therefore to use probiotics is that in such patients, administration of exogenous commensal microorganisms (that is probiotics) is required to restore the microflora to one that more closely reflects the normal flora prior to antibiotic therapy. Some open ended studies have indeed shown that this approach can alleviate the signs and symptoms of C. difficile infection. With respect to antibiotic-associated diarrhea, probiotics have proved useful as a prophylactic regimen, and potentially they can also be used to alleviate the signs and symptoms once antibiotic induced diarrhea has occurred. It must be recognized that evidence for therapeutic effects against C. difficile, and other disorders has been obtained using certain probiotic strains, such as L. rhamnosus GG.

Helicobacter pylori infection and complications

A new development for probiotic applications is activity against Helicobacter pylori, a Gram negative pathogen responsible for type B gastritis, peptic ulcers and gastric cancer. In vitro and animal data indicate that lactic acid bacteria can inhibit the growth of the pathogen and decrease urease enzyme activity necessary for the pathogen to remain in the acidic environment of the stomach. Human data is limited, but there is some evidence of an effect induced by L. johnsonii La1. In terms of measuring probiotic effects, feasible end points include the suppression of the infection (which may be reversible upon cessation of treatment), combination treatment with antibiotics leading to fewer side effects such as acid reflux, and lower risk of recurrent infection. Placebo-controlled trials are needed before specific claims can be made for probiotic anti-Helicobacter pylori benefits in humans with respect to prevention and treatment. Such studies are warranted given the preliminary evidence to support these effects.

Inflammatory diseases and bowel syndromes

Inflammatory bowel diseases, such as pouchitis and Crohn's disease, as well as irritable bowel syndrome, may be caused or aggravated by alterations in the gut flora including infection. These are new avenues of investigation, although it is premature to state a firm action of probiotics in these conditions. Some studies support the potential role of probiotics in therapy and prophylaxis and illustrate that combinations of strains may have a role to play in remediation. The intestinal microflora likely plays a critical role in inflammatory conditions in the gut, and potentially probiotics could remediate such conditions through modulation of the microflora. Clinical and mechanistic studies are urgently required to better understand the interface between the microbes, host cells, mucus and immune defenses, and to createefficacious interventions. Such studies should include molecular examination of the intestinal (not only fecal) flora and long-term (5-10 years) effects of probiotic microorganisms.

Cancer

There is some preliminary evidence that probiotic microorganisms can prevent or delay the onset of certain cancers. This stems from the knowledge that members of the gut microflora can produce carcinogens such as nitrosamines. Therefore, administration of lactobacilli and bifidobacteria could theoretically modify the flora leading to decreased â-glucuronidase and carcinogen levels. Furthermore, there is some evidence that cancer recurrences at other sites, such as the urinary bladder can be reduced by intestinal instillation of probiotics including L. casei Shirota. In vitro studies with L. rhamnosus GG and bifidobacteria and an in vivo study using L. rhamnosus strains GG and LC-705 as well as Propionibacterium sp. showed a decrease in availability of carcinogenic aflatoxin in the lumen. However, it is too early to make definitive clinical conclusions regarding the efficacy of probiotics in cancer prevention.

There is sufficient proof of a correlation between probiotics and specific anti-cancer effects, and urged that extensive studies are required. Such studies must utilize internationally recognized markers for cancer, or risk of cancer, and evaluate such markers and presence of carcinogenic lesions or tumors over a suitably long period of time for prevention of primary cancer, and reduction of the incidence of recurrences.

Constipation

The ability of probiotic therapy to alleviate constipation (difficulty in passing stool, excessive hardness of stool, slow transit through the bowel) is debatable, but may be a feature of selected strains. Randomized placebo controlled efficacy studies aimed at exploring these effects are strongly recommended.

Mucosal Immunity

The innate and adaptive immune systems are the two compartments traditionally described as important for the immune response. Macrophages, neutrophils, natural killer (NK) cells and serum complement represent the main components of the innate system, in charge of the first line of defence against many microorganisms. However, there are many agents that this system is unable to recognize. The adaptive system (B and T cells) provide additional means of defence, while cells of the innate system modulate the beginning and subsequent direction of adaptive immune responses. Natural killer cells, including gamma/delta T cells, regulate the development of allergic airway disease, suggesting that the interleukins play an important role. Intravenous, intraperitoneal and intrapleural injection of L. casei Shirota into mice significantly increased NK activity of mesenteric node cells but not of Peyer's patch cells or of spleen cells, supporting the concept that some probiotic strains can enhance the innate immune response.

A number of studies have been performed in vitro and in animals which clearly show that probiotic strains can modify immune parameters. Correlating these findings with events taking place in the human body is still somewhat unclear, but evidence is mounting that such effects occur. In a series of randomized, double blind, placebo controlled clinical trials, it was demonstrated that dietary consumption of B. lactis HN019 and L. rhamnosus HN001 resulted in measurable enhancement of immune parameters in the elderly.

Probiotic modulation of host immunity is a very promising area for research. Supportive data is emerging, such as those carried out in humans showing that probiotic microorganisms can enhance NK cell activity in the elderly and non-specific host defenses can be modulated.

There is a need to specify whether the activities being advocated are designed to operate in otherwise healthy people or subjects with known diseases. Some of the critical factors involved in the host's defenses have been identified and include the induction of mucus production or macrophage activation by lactobacilli signaling, stimulation of sIgA and neutrophils at the site of probiotic action (for example the gut), and lack of release of inflammatory cytokines or stimulation of elevated peripheral immunoglobulins. It is also recognized that in some situations, stimulation of factors such as inflammatory cytokines may confer health benefits on the host.

Future studies should focus on the effect in humans, and elucidate the mechanisms of action within systems which simulate the in vivo situation, and link this to bacterial and human genomics.

Allergy

In a double-blind, randomized, placebo-controlled trial, L. rhamnosus GG was given to pregnant women for four weeks prior to delivery, then to newborns at high risk of allergy for six months with the result that there was a significant reduction in early atopic disease. This study illustrates the potential for probiotic microorganisms to modulate the immune response and prevent onset of allergic diseases. In other clinical studies with infants allergic to cow's milk, atopic dermatitis was alleviated by ingestion of probiotic strains L. rhamnosus GG and B. lactis BB-12. The precise mechanisms have not been elucidated, but the premise is based upon the ability of lactobacilli to reverse increased intestinal permeability, enhance gut-specific IgA responses, promote gut barrier function through restoration of normal microbes, and enhance transforming growth factor beta and interleukin 10 production as well as cytokines that promote production of IgE antibodies. Whether T-helper-1 (TH1) is enhanced and/or T-helper-2 (TH2) dominance is reduced remains to be determined, as do the time-points of these types of events. Certain microorganisms can contribute to the generation of counter-regulatory T-helper cell immune responses, indicating that use of specific probiotic microorganisms could redirect the polarized immunological memory to a healthy one.

Cardiovascular Disease

There is preliminary evidence that use of probiotic lactobacilli and metabolic by-products potentially confer benefits to the heart, including prevention and therapy of various ischemic heart syndromes and lowering serum cholesterol.

Urogenital Tract Disorders

Excluding sexually transmitted diseases, almost all infections of the vagina and bladder are caused by microorganisms that originate in the bowel. There is a strong correlation between presence of commensals, particularly lactobacilli in the vagina with health, and an absence of these microorganisms in patients with urogenital infections. Disruption of the normal vaginal flora is caused by broad-spectrum antibiotics, spermicides, hormones, dietary substances and factors not, as yet, fully understood. There is some evidence that probiotic microorganisms delivered as foods and topical preparations have a role in preventing urogenital tract disorders. The criteria for selection of effective probiotic strains have been proposed and should include verification of safety, colonization ability in the vagina and ability to reduce the pathogen count through competitive exclusion of adherence and inhibition of pathogen growth.

Bacterial vaginosis

Bacterial vaginosis (BV) is a disease of unknown etiology resulting from the overgrowth of various anaerobic bacterial species and associated with the disappearance of lactobacilli, which dominate the normal vagina. Many women with BV are asymptomatic yet are at risk of more serious complications such as endometriosis, pelvic inflammatory disease and complications of pregnancy including pre-term labour. There is some clinical evidence to suggest that oral and vaginal administration of lactobacilli can eradicate asymptomatic and symptomatic BV. Oral administration of Lactobacillus acidophilus and yogurt has been used in the prevention and therapy of candidal vaginitis, although no efficacy data have yet been generated. The necessity for the lactobacilli to produce hydrogen peroxide has been proposed, but given that these microorganisms are more prone to being killed by spermicides, the combination of two or more strains, one of which produces hydrogen peroxide and others which resists spermicidal killing, may prove to be more therapeutic.

Yeast vaginitis

Yeast vaginitis is a very common ailment, often precipitated by antibiotic use, exposure to spermicides or hormonal changes as yet not fully understood. Unlike BV and urinary tract infection, yeast vaginitis is not necessarily due to loss of lactobacilli. Few Lactobacillus strains are able to inhibit the growth and adhesion of Candida albicans or other Candida species, and there is no solid evidence to indicate that intravaginal administration of lactobacilli can eradicate yeast infection. However, there is some evidence to suggest that lactobacilli ingestion and vaginal use can reduce the risk of recurrences and further studies are warranted since this disease is widespread and debilitating.

Urinary tract infections

Several hundred million women are affected by urinary tract infection (UTI) annually. Uropathogenic Escherichia coli originating in the bowel is the responsible agent in up to 85% of cases. Asymptomatic bacteruria is also a common finding in women, and sometimes it is followed by symptomatic UTI. There is evidence, including randomized controlled data to suggest that once weekly vaginal capsules of freeze dried Lactobacillus strains GR-1 and B-54 prepared with addition of skim milk, and once daily oral capsule use of Lactobacillus strains GR-1 and RC-14, can result in the restoration of a lactobacilli dominated vaginal flora and lower risk of UTI recurrences. By creating a lactobacilli barrier in the vagina, it is believed that fewer pathogens can ascend into the bladder, thereby blocking the infectious process.

Use of probiotics in otherwise healthy people

Many probiotic products are used by consumers who regard themselves as being otherwise healthy. They do so on the assumption that probiotics can retain their health and well being, and potentially reduce their long-term risk of diseases of the bowel, kidney, respiratory tract and heart. Several points need to be made on this assumption and its implications. The use of probiotics should not replace a healthy lifestyle and balanced diet in otherwise healthy people.

Firstly, there is no precise measure of "health" and subjects may actually have underlying and undetectable diseases at any given time. Secondly, no studies have yet been undertaken which analyse whether or not probiotic intake on a regular basis helps retain life-long "health" over and above dietary, exercise and other lifestyle measures. One study of day care centres in Finland showed that probiotic use reduced the incidence of respiratory infections and days absent due to ill health.

It is currently unclear as to the impact of regular probiotic intake on the intestinal microflora. For example, does it lead to the depletion or loss of commensal microorganisms which otherwise have beneficial effects on the host? While there is no indication of such effects, the issue needs to be considered. Furthermore, the concept of restoring a normal balance assumes that we know what the normal situation in any given intestinal tract comprises. The ingestion of probiotic strains has not led to measurable long-term colonization and survival in the host. Invariably, the microorganisms are retained for days or weeks, but no longer. Thus, use of probiotics likely confers more transient than long-term effects, and so continued intake appears to be required.

In newborn children, where a commensal flora has not yet been established, it is feasible that probiotic microorganisms could become primary colonizers that remain long-term, perhaps even for life. While such probiotic usage can prevent death and serious morbidity in premature, low birth weight infants, the alteration of flora in healthy babies is a more complex situation. Just so, an implication of the Human Genome Project is that selected probiotics may be used at birth to create a flora that improves life-long health. These issues are very important for the future, and will require full discussion including human ethical considerations.

Testing Methods for Health Benefits of Probiotics

Proper in vitro studies should establish the potential health benefits of probiotics prior to undertaking in vivo trials. Tests such as acid and bile tolerance, antimicrobial production and adherence ability to human intestinal cells should be performed depending on the proposed health benefit.

In order to ascertain that a given probiotic can prevent or treat a specific pathogen infection, a clinical study must be designed to verify exposure to the said pathogen (preventive study), or that the infecting microorganism is that specific pathogen (treatment study). If the goal is to apply probiotics in general to prevent or treat a number of infectious gastroenteritis or urogenital conditions, the study design must define the clinical presentation, symptoms and signs of infection, and include appropriate controls.

For in vivo testing, randomized double blind, placebo controlled human trials should be undertaken to establish the efficacy of the probiotic product. There is a need for human studies in which adequate numbers of subjects are enrolled to achieve statistical significance. It would be preferable to have such findings corroborated by more than one independent center. For some foods, it may be difficult to separate a probiotic effect from an effect related to the general product characteristics of the food. Therefore, it is essential that proper controls be included in these human trials. Furthermore, data obtained with one specific probiotic food cannot be extrapolated to other foods containing that particular probiotic strain or to other probiotic microorganisms.

With respect to measuring the health benefits in human studies, consideration should be given to clinically relevant outcomes in the population being studied. For diarrheal studies, this might be preventing death in some countries, while in others it might be prevention of a defined and statistically significant weight loss, decreased duration of watery/liquid stools, and faster recovery to normal health, as measured by restoration of normal bowel function and stool consistency.

Although it is known that certain probiotics can elicit beneficial effects, little is known about the molecular mechanisms of the benefits reported. The mechanisms may vary from one probiotic to another (for the same benefit via different means) and the mechanism may be a combination of events, thus making this a very difficult and complex area. It could involve the production of a specific enzyme(s) or metabolite(s) that act directly on the microorganism(s), or the probiotic could also cause the body to produce the beneficial action.

Examples of possible probiotic mechanisms of action, in the control of intestinal pathogens include:

- Antimicrobial substance production
- Competitive exclusion of pathogen binding
- Competition for nutrients
- Modulation of the immune system

Clear experiments (in vitro and/or in vivo) should be designed at the molecular level to elucidate the mechanisms of probiotic beneficial effects. Appropriate experiments including genetic analysis to elucidate the mechanism of actions should be performed.

Probiotic bacteria containing â-galactosidase can be added to food to improve lactose maldigestion. However, a similar health effect is also observed for lactose fermenting starter bacteria such as L. delbrueckii. ssp. bulgaricus and S. thermophilus in fermented milk products like yogurt. These traditional starters are not considered probiotics since they lack the ability to proliferate in the intestine.

Safety Considerations

Antimicrobial Resistance Profiles of Probiotics

As with any bacteria, antibiotic resistance exists among some lactic acid bacteria, including probiotic microorganisms. This resistance may be related to chromosomal, transposon or plasmid located genes. However, insufficient information is available on situations in which these genetic elements could be mobilized and it is not known if situations could arise where this would become a clinical problem.

There is concern over the use in foods of probiotic bacteria that contain specific drug resistance genes. Bacteria, which contain transmissible drug resistance genes, should not be used in foods. Currently, no standardized phenotypic methods are available which are internationally recognized for lactobacilli and bifidobacteria (non-pathogens).

Plasmids exist in lactobacilli and bifidobacteria, especially in strains isolated from the intestine, which have genes encoding antibiotic resistance. Due to the relevance of this problem, it is suggested that further research be done relating to the antibiotic resistance of lactobacilli and bifidobacteria.

When dealing with selection of probiotic strains, it is recommended that probiotic bacteria should not harbour transmissible drug resistance genes encoding resistance to clinically used drugs. Research is required relating to the antibiotic resistance of lactobacilli and bifidobacteria and the potential for transmission of genetic elements to other intestinal and/or foodborne microorganisms.

Safety of Probiotics in Humans

In terms of safety of probiotics, a set of general principles and practical criteria need to be generated to provide guidelines as to how any given potential probiotic microorganism can be tested and proven to have a low risk of inducing or being associated with the etiology of disease, versus conferring a significant health benefit when administered to humans. These guidelines should recognize that some species may require more vigorous assessment than others. In this respect, the evaluation of safety will require at least some studies to be performed in humans, and should address aspects of the proposed end use of the probiotic strain.

Information acquired to date shows that lactobacilli have a long history of use as probiotics without established risk to humans, and this remains the best proof of their safety. Also, no pathogenic or virulence properties have been found for lactobacilli, bifidobacteria or lactococci. Under certain conditions, some lactobacilli strains have been associated with adverse effects, such as rare cases of bacteremia. However, a recent epidemiological study of systematically collected lactobacilli bacteremia case reports in one country has shown that there is no increased incidence or frequency of bacteremia with increased usage of probiotic lactobacilli.

It is also acknowledged that some members of lactic acid bacteria, such as enterococci may possess virulence characteristics. For this and other reasons, Enterococcus not be referred to as a probiotic for human use. The rationale is based upon:

a. Strains can display a high level of resistance to vancomycin, or can acquire such resistance.

If this resistance is present, transfer to other microorganisms may occur and thiscould enhance the pathogenesis of such recipients

b. Certain strains of vancomycin resistant enterococci are commonly associated with nosocomial infections in hospitals.

Some strains of Enterococcus display probiotic properties, and may not at the point of inclusion in a product display vancomycin resistance. However, the onus is on the producer to prove that any given strain cannot acquire or transfer vancomycin resistance or be virulent and induce infection.

Quality Assurance and Regulatory Issues

Regulatory Issues

Government regulations differ among countries, however the status of probiotics as a component in food is currently not established on an international basis. For the most part, probiotics come under food and dietary supplements because most are delivered by mouth as foods. These are differentiated from drugs in a number of ways, especially with respect to claims. Drugs are allowed to claim effectiveness in the treatment, mitigation or cure of a disease, whereas foods, feed additives and dietary supplements can only make general health claims.

In order to understand where probiotic products currently fall in terms of regulatory agencies, and the claims that can be made with their use, the following US example is provided. Consumers are permitted access to products ingested as pills, capsules, tablets and liquids, or in capsules sold in health food stores or via the internet.

— A 'health claim' is defined as "a statement, which characterizes the relationship of any substance to a disease or health-related condition, and these should be based upon well-established, generally accepted knowledge from evidence in the scientific literature and/or recommendations from national or international public health bodies. Examples include 'protects against cancer'.

— A structure/function claim is defined as "a statement of nutritional support that describes the role of a nutrient or dietary ingredient to affect the structure or functioning of the human body, or characterizes the documented mechanism by which a nutrient or dietary ingredient acts to maintain such structure or function. Examples include 'supports the immune system'. Claims that substances can treat, diagnose, cure or prevent a disease are not structure/function claims.

The new paradigm of risk analysis is making its way into regulatory food safety and focuses on a functional separation of the science-based risk assessment and risk management. However, the issue of communication is now also considered an important integrated part of risk analysis. Communication includes exchange between assessors and managers and two-way interaction with other interested parties. Within this concept, the transparency of the decision making process for food safety regulatory action is emphasized, as well as the importance of providing a vehicle for consumers and others to participate in the development process. Therefore communication efforts relative to the use of probiotics should be considered as an integrated part of the development of regulatory initiatives.

Appropriate Labelling

To clarify the identity of a probiotic present in the food, the microbial species be stated on the label. If a selection process has been undertaken at the strain level, the identity of the strain should also be included, since the probiotic effect seems to be strain specific.

There is a need to accurately enumerate the probiotic bacteria in food products in order to include them on the label. The label should state the viable concentration of each probiotic present at the end of shelf life.

Manufacturing and Handling procedures

To ensure that any given culture maintains the beneficial properties, the stock culture should be maintained under appropriate conditions and be checked periodically for strain identity and probiotic properties. Furthermore, viability and probiotic activity must be maintained throughout processing, handling and storage of the food product containing the probiotic, and verified at the end of shelf-life.

Powdered milk products

There is not adequate information available on the stability of probiotics in powdered milk and little information is available on the issue of probiotic quality following spray drying. Cell damage and loss of viability of the probiotic culture occur during the spray drying process. Thus improvements in spray drying methods are necessary to ensure better survival, including the use of protective agents which have been shown to enhance survival of lactobacilli and environmental adaptation. Probiotic stability during powder storage is inversely related to storage temperature, and methods have to be identified to address this. Although not published in the literature, certain companies producing starter cultures have the technology to produce freeze dried lactic acid bacteria including probiotics that are 'stabilized' and thus retain a high level of viability during drying and storage. The incorporation of such dried cultures into powdered milk may be the method of choice for preparing powdered milk products containing probiotics. However, research is needed including storage testing to confirm the feasibility of such a process.

Careful consideration should be given to factors such as the following, with respect to viability of the probiotic:

- Drying method
- Type of packaging
- Size of packaging
- Storage conditions (temperature, humidity, etc.)
- Powder milk quality (Standard reference)
- Rehydration procedure
- Handling of rehydrated product

Prebiotics

Prebiotics are generally defined as 'nondigestible food ingredients that beneficially affect the host by selectively stimulating the growth and/or activity of one or a limited number of bacterial species already established in the colon, and thus in effect improve host health'.

The concept of prebiotics essentially has the same aim as probiotics, which is to improve host health via modulation of the intestinal flora, although by a different mechanism. However, there are some cases in which prebiotics may be beneficial for the probiotic, especially with regard to bifidobacteria, that is the synbiotic concept. Synbiotics are defined as 'mixtures of probiotics and prebiotics that beneficially affect the host by improving the survival and implantation of live microbial dietary supplements in the gastrointestinal tract of the host'.

Post Market Surveillance

Probiotic producers, medical professionals and public health officers consider some form of system to monitor the health outcomes of long-term probiotic administration. This is suggested as a means to gain insight into potential side effects as well as assess long-term benefits. A necessary prerequisite for surveillance is a proper trace-back system.

The experts agreed that adequate scientific evidence exists to indicate that there is potential for the derivation of health benefits from consuming food containing probiotics There is good evidence that specific strains of probiotics are safe for human use and able to confer some health benefits on the host, but such benefits cannot be extrapolated to other strains without experimentation.

The health benefits for which probiotics can be applied include conditions such as gastrointestinal infections, certain bowel disorders, allergy, and urogenital infections, which afflict a large portion of the world's population. The application of probiotics to prevent and treat these disorders should be more widely considered by the medical community. In addition, there is emerging evidence to indicate that probiotics can be taken by otherwise healthy people as a means to prevent certain diseases and modulate host immunity.

The regulatory status of probiotics as a component in food is currently not established on an international basis. In only a few countries, regulatory procedures are in place or sufficiently developed to enable probiotic products to be allowed to describe specific health benefits.

Potential probiotic strains must be identified by methods including internationally accepted molecular techniques and named according to the International Code of Nomenclature, and strains should preferably be deposited in a reputable internationally recognized culture collection.

There is a need for refinement of in vitro and in vivo tests to better predict the ability of probiotic microorganisms to function in humans.

Good manufacturing practices must be applied with quality assurance, and shelf-life conditions established, and labeling made clear to include minimum dosage and verifiable health claims.

The regulatory status of probiotics as a component in food has to be established on an international level. A regulatory framework be established to better address issues related to probiotics including efficacy, safety, labelling, fraud and claims.

Probiotic products shown to confer defined health benefits on the host should be permitted to describe these specific health benefits.

Surveillance systems, including trace-back and post marketing surveillance, should be put in place to record and analyze any adverse events associated with probiotics in food. Such systems could also be used to monitor the long-term health benefits of probiotic strains.

Efforts should be made to make probiotic products more widely available, especially for relief work and populations at high risk of morbidity and mortality.

Guidelines for the Evaluation of Probiotics in Food

The Joint FAO/WHO Expert Consultation on Evaluation of Health and Nutritional Properties of Probiotics in Food held in Córdoba, Argentina from 1 to 4 October, 2001 recognized that there is a need for guidelines to set out a systematic approach for the evaluation of probiotics in food leading to the substantiation of health claims. Consequently, a Working Group was convened by FAO/WHO to generate guidelines and recommend criteria and methodology for the evaluation of probiotics, and to identify and define what data need to be available to accurately substantiate health claims. The aims of the Working Group were to identify and outline the minimum requirements needed for probiotic status. Consequently, guidelines were prepared to meet this objective.

Guidelines for Probiotics

Genus/species/strain

It was recognized that it is necessary to know the genus and species of the probiotic strain. The current state of evidence suggests that probiotic effects are strain specific. Strain identity is important to link a strain to a specific health effect as well as to enable accurate surveillance and epidemiological studies. A possible exception is the ability in general of S. themophilus and L. delbrueckii ssp. bulgaricus to enhance lactose digestion in lactose intolerant individuals. In this case, or in other cases where there is suitable scientific substantiation of health benefits that are not strain specific, individual strain identity is not critical.

Speciation of the bacteria must be established using the most current, valid methodology. It is recommended that a combination of phenotypic and genetic tests be used.

Nomenclature of the bacteria must conform to the current, scientifically recognized names. Protracted use of older or misleading nomenclature is not acceptable on product labels. The use of incorrect names does not properly identify the probiotic bacterium in the product and forces consumers and regulatory agencies to make assumptions about the identity of the real bacterium being sold.

DNA-DNA hybridization is the reference method to specify that a strain belongs to a species; however, as it is time consuming and beyond the resources of many laboratories, requiring a large collection of reference strains, the use of DNA sequences encoding 16S rRNA is suggested as a suitable substitute. In this case, it is recommended that this genotypic technique be combined with phenotypic tests for confirmation.

Patterns generated from the fermentation of a range of sugars and final fermentation products obtained from glucose utilization are key phenotypes that should be investigated for identification purposes.

Strain typing has to be performed with a reproducible genetic method or using a unique phenotypic trait. Pulsed Field Gel Electrophoresis (PFGE) is the gold standard. Randomly Amplified Polymorphic DNA (RAPD) can also be used, but is less reproducible. Determination of the presence of extrachromosomal genetic elements, such as plasmids can contribute to strain typing and characterization. It is recommended that all strains be deposited in an internationally recognized culture collection.

In vitro tests to screen potential probiotics

In vitro tests are critical to assess the safety of probiotic microbes.

In addition, in vitro tests are useful to gain knowledge of strains and the mechanism of the probiotic effect. However, it was noted that the currently available tests are not fully adequate to predict the functionality of probiotic microorganisms in the human body. It was also noted that in vitro data available for particular strains are not sufficient for describing them as probiotic. Probiotics for human use will require substantiation of efficacy with human trials. Appropriate target-specific in vitro tests that correlate with in vivo results are recommended. For example, in vitro bile salts resistance was shown to correlate with gastric survival in vivo. A list of the main currently in vitro tests used for the study of probiotic strains is shown in Table 1. All of these tests require validation, however, with in vivo performance.

Table 1. Main currently used in vitro tests for the study of probiotic strains

Resistance to gastric acidity
Bile acid resistance
Adherence to mucus and/or human epithelial cells and cell lines
Antimicrobial activity against potentially pathogenic bacteria
Ability to reduce pathogen adhesion to surfaces
Bile salt hydrolase activity
Resistance to spermicides (applicable to probiotics for vaginal use)

Safety Considerations

Historically, lactobacilli and bifidobacteria associated with food have been considered to be safe. Their occurrence as normal commensals of the mammalian flora and their established safe use in a diversity of foods and supplement products worldwide supports this conclusion. However, probiotics may theoretically be responsible for four types of side-effects :

1. Systemic infections
2. Deleterious metabolic activities
3. Excessive immune stimulation in susceptible individuals
4. Gene transfer

Documented correlations between systemic infections and probiotic consumption are few and all occurred in patients with underlying medical conditions. The following is a list (including some microbes used in non-food applications) of infections reported to be associated (although not necessarily proven) with the consumption of commercial products:

- Two cases of L. rhamnosus traced to possible probiotic consumption.
- Thirteen cases of Saccharomyces fungemia due to vascular catheter contamination.
- Bacillus infections linked to probiotic consumption include three reports detailing seven cases of B. subtilis bacteremia, septicemia and cholangitis, all in patients with underlying disease.

No cases of infections from Bifidobacterium have been reported. Enterococcus is emerging as an important cause of nosocomial infections and isolates are increasingly vancomycin resistant. The Working Group recognizes that some strains of Enterococcus display probiotic properties, and may not at the point of inclusion in a product display vancomycin resistance. However, the onus is on the producer to prove that any given probiotic strain is not a significant risk with regard to transferable antibiotic resistance or other opportunistic virulence properties.

In recognition of the importance of assuring safety, even among a group of bacteria that is Generally Recognized as Safe (GRAS), the Working Group recommends that probiotic strains be characterized at a minimum with the following tests:

1. Determination of antibiotic resistance patterns;
2. Assessment of certain metabolic activities (e.g., D-lactate production, bile salt deconjugation);
3. Assessment of side-effects during human studies;
4. Epidemiological surveillance of adverse incidents in consumers (post-market);
5. If the strain under evaluation belongs to a species that is a known mammalian toxin producer, it must be tested for toxin production. One possible scheme for testing toxin production has been recommended by the EU Scientific Committee on Animal Nutrition ;
6. If the strain under evaluation belongs to a species with known hemolytic potential, determination of hemolytic activity is required.

Assessment of lack of infectivity by a probiotic strain in immunocompromized animals would add a measure of confidence in the safety of the probiotic.

In vivo Studies Using Animals and Humans

In some cases, animal models exist to provide substantiation of in vitro effects and determination of probiotic mechanism. Where appropriate, the Working Group encourages use of these prior to human trials.

The principal outcome of efficacy studies on probiotics should be proven benefits in human trials, such as statistically and biologically significant improvement in condition, symptoms, signs, well-being or quality of life; reduced risk of disease or longer time to next occurrence; or faster recovery from illness. Each should have a proven correlation with the probiotic tested.

Probiotics have been tested for an impact on a variety of clinical conditions. Standard methods for clinical evaluations are comprised of Phase 1 (safety), Phase 2 (efficacy), Phase 3 (effectiveness) and Phase 4 (surveillance).

Probiotics delivered in food generally are not tested in Phase 3 studies, which are concerned with comparison with a standard therapy. When a claim is made for a probiotic altering a disease state, the claim should be made based on sound scientific evidence in human subjects.

In Phase 2 and 3 studies, the Working Group recognizes the value of validated quality of life assessment tools.It is recommended that human trials be repeated by more than one Center for confirmation of results.

No adverse effects related to probiotic administration should be experienced when food is considered. Adverse effects should be monitored and incidents reported.

Information accumulated to show that a strain(s) is a probiotic, including clinical trial evidence be published in peer-reviewed scientific or medical journals. Furthermore, publication of negative results is encouraged as these contribute to the totality of the evidence to support probiotic efficacy.

References

Adams MR, Marteau P. 1995. *On the safety of lactic acid bacteria.* Int J Food Micro, 27: 263-264.

Conway PL, Gorbach SL, Goldin BR. 1987. *Survival of lactic acid bacteria in the human stomach and adhesion to intestinal cells.* J Dairy Sci, 70: 1-12.

Hennequin C, Kauffmann-Lacroix C, Jobert A, Viard JP, Ricour C, Jacquemin JL, Berche P. 2000. *Possible role of catheters in Saccharomyces boulardii fungemia.* Eur J Clin Microbiol Infect Dis, 19: 16-20.

Oggioni MR, Pozzi G, Balensin PE, Galieni P, Bigazzi C . 1998. *Recurrent* septicemia in an immunocompromised patient due to probiotic strains of Bacillus *subtilis.* J Clin Microbiol, 36: 325-326.

SCAN. 2000. *Report of the Scientific Committee on Animal Nutrition on the Safety of Use of Bacillus Species in Animal Nutrition.* European Commission

7

Microbiology of Animal Feeds

A wide range of microbes occurs naturally on, or as contaminants of forages, cereal grains, oilseed by-products and compound feeds. Beneficial effects can result from the occurrence of the lactic acid bacteria during the fermentation of forages during the process of ensilage. These bacteria favour the production of lactic acid, which helps in reducing the pH to around 4.0, thus preserving the forage for winter-feeding to ruminants. Lactic acid bacteria and yeast cultures have also been attributed with beneficial properties as feed probiotics for reducing scouring and increasing growth performance in farm animals. Animal feeds may become contaminated with harmful bacteria such as Salmonella, Listeria and E. coli. In the case of the latter organism, faecal sources and slurry have been identified as the primary routes of contamination both in pastures and compound feeds. Cereal grains and oilseed by-products are regularly contaminated with fungi occurring as plant pathogens or developing during storage. Major adverse effects arise in farm animals due to the production of mycotoxins by certain species and strains of these fungi. The use of fungicides and preservatives represents potential methods for reducing the prevalence of deleterious fungi. Extensive legislation exists for the control of a number of feed contaminants, particularly mycotoxins.

The microbiology of animal feeds emerged as an important issue in the wake of the *Salmonella*, *E. coli* O157,*Campylobacter* and BSE crises in the EU and elsewhere and the momentum has been maintained by the foot and mouth epidemic of 2001/2 in the United Kingdom. Fungal contamination of animal feeds is a regular occurrence on a worldwide scale and detrimental effects have been observed in all classes of farm animals due to the production of mycotoxins by certain species and strains of moulds. There are few grounds for complacency as regards animal feed safety and vigilance should now be the watchword for all those involved in the livestock industry. Nevertheless, it should be recognised that successful silage production depends upon the promotion of fermentation brought about by beneficial bacteria.

Silage Microbiology

Successful preservation of high-moisture forage and other crops depends upon the controlling the activities of microbes, particularly bacteria. Conditions favouring optimal preservation include the rapid imposition and subsequent maintenance of anaerobic conditions at all stages in the process of ensilage. Under these conditions, lactic acid bacteria proliferate, using endogenous plant sugars to produce sufficient quantities of the acid to depress pH to around 4, the optimum value for successful preservation. Wet forages are difficult to preserve in this way and often provide ideal conditions for the growth of undesirable bacteria such as the Clostridia.

Species

The lactic acid bacteria (LAB) are of two types. The homofermentative species include *Lactobacillus plantarum,Pediococcus pentosaceus* and *Enterococcus faecalis*. The heterofermentative species comprise *Lactobacillus brevis* and*Leuconostoc mesenteroides*. The homofermentative group are more efficient in converting forage sugars into lactic acid. Of these species *L. plantarum* is generally considered to be the most competitive, rapidly producing large amounts of lactic acid in freshly ensiled forage. LAB are essentially non-proteolytic organisms and are therefore considered to be beneficial in that they assist in conserving labile proteins and free amino acids in forage.

Two groups of Clostridia are also recognised. The saccharolytic group includes *Clostridium butyricum* and *C. tyrobutyricum*. These species ferment residual sugars as well as lactic acid to butyric acid, causing a rise in pH. The proteolytic group includes *C. bifermentans* and *C. sporogenes*. These bacteria ferment amino acids to a number of different products. The Clostridia thrive under high-moisture conditions and consequently the ensilage of wet forages is fraught with problems associated with adequate preservation and satisfactory voluntary intakes and performance of animals fed on such silages.

The Enterobacteria include *Escherichia coli* and *Erwinia herbicola* and are also considered to be undesirable in that they compete with LAB for plant sugars fermenting them to acetic acid, ethanol, CO_2 and H_2. They are also capable of catabolising amino acids to NH_3.

Listeria monocytogenes is widely distributed in nature and also occurs in silage, particularly big bale silage. The increased incidence of listeriosis in sheep and cattle has been partly linked with the introduction of big bale silage. The relatively low density and limited fermentation characteristics of big bale silage and susceptibility of the bags to damage all favour the growth of *L. monocytogenes*. This organism is of particular significance because of it potential to contaminate animal products destined for human consumption.

Fungi occur in silages as yeasts and moulds. The yeasts include species of *Candida*, *Saccharomyces* and *Torulopsis*. The moulds associated with silages include various species of *Aspergillus*, *Penicillium* and *Fusarium*. The occurrence of these fungi is of particular concern due to their potential to produce harmful mycotoxins.

Modulation of Microbial Activity

The activities of bacteria may be controlled to facilitate preservation of ensiled forages. Wilting the crop before ensiling is a common method of restricting fermentation, allowing the growth of LAB but inhibiting the activities of undesirable organisms such as Clostridia and Enterobacteria. An added benefit of wilting is that effluent production is reduced, thus minimising environmental pollution from this source.

Additives may be used to control or stimulate fermentation by bacteria in ensiled forages. Formic acid is widely used to restrict fermentation by artificially reducing pH to below 4.0. Commercial products are available containing, for example, 85*per cent* formic acid which is applied during the harvesting of the forage. However, even at the recommended rates of application, the bacteria are still able to ferment sugars to lactic acid. Stimulation of silage bacteria may be brought about in two ways. Addition of molasses to the forage prior to ensiling ensures sufficient substrate for LAB thus causing the desired reduction in pH values. More recently, however, it has been suggested that a limiting factor is the relatively low concentrations of LAB in forage crops at harvest. Consequently, a number of commercial inoculants containing freeze-dried cultures of homofermentative LAB have become available for application to harvested forage. Efficacy of this procedure depends upon the inoculation rate as well as the presence of adequate concentrations of sugars.

Fungal Contamination of Concentrates and Forages

Fungal contamination of cereal grains, oil-seed meals and forages continues to represent a major animal health risk throughout the world and particularly in the humid tropics. The risks arise primarily from the ability of particular species and strains of fungi to produce harmful compounds known as mycotoxins. These substances arise from the secondary metabolism of fungi in response to a wide range of genetic and environmental factors.

Mycotoxin contamination of forages and cereals frequently occurs in the field following infection of plants with specific pathogenic fungi or with symbiotic endophytes. In addition, contamination may occur during processing and storage of harvested products and feed whenever environmental conditions are appropriate for spoilage fungi. Moisture content and ambient temperature are key factors affecting fungal colonisation of and mycotoxin production in concentrates and compound feeds.

Inhalation of fungal spores or consumption of mycelia may cause conditions collectively known as 'mycoses'. An example of such a condition is mycotic abortion arising from the systemic transmission and delivery of fungal material to the placenta.

Classification

It is conventional to subdivide toxigenic fungi into 'field' (or plant pathogenic) and 'storage' (or saprophytic/spoilage) organisms. *Claviceps*, *Neotyphodium Fusarium* and *Alternaria* are classical representatives of field fungi while *Aspergillus*and *Penicillium* exemplify storage organisms. Mycotoxigenic species may be further distinguished on the basis of geographical prevalence, reflecting specific environmental requirements for growth and secondary metabolism. Thus,*Aspergillus flavus*, *Aspergillus parasiticus* and *Aspergillus ochraceus* readily proliferate under warm, humid conditions whereas *P. expansum* and *P. verrucosum* are essentially temperate fungi. Consequently, the *Aspergillus* mycotoxins predominate in plant products emanating from the tropics and other warm regions, while the *Penicillium* mycotoxins occur widely in temperate foods, particularly cereal grains. *Fusarium* fungi are more ubiquitous, but even this genus contains toxigenic species which are almost exclusively associated with cereals from tropical and sub-tropical countries.

Aspergillus Species

The *Aspergillus* genus dominates all other fungi in respect of mycotoxin production in cereals and oilseeds. For example, Dhand *et al.* (1998) observed that *Aspergillus* was the most significant genus in dairy and other feeds in the tropics. Three species are responsible for virtually all mycotoxin production by this genus: *Aspergillus flavus*, *Aspergillus parasiticus* and*Aspergillus ochraceus*. *A. flavus*, and *A. parasiticus* synthesise the aflatoxins, while *A. ochraceus* produces the ochratoxins.

The aflatoxins include aflatoxin B_1, B_2, G_1 and G_2 (AFB_1, AFB_2, AFG_1 and AFG_2, respectively). In addition, aflatoxin M_1(AFM_1) may well occur in the milk of dairy cows consuming AFB_1-contaminated feeds. The aflatoxigenic *Aspergilli* are generally regarded as storage fungi, proliferating under conditions of relatively high moisture / humidity and temperature. Aflatoxin contamination is, therefore, almost exclusively confined to tropical feeds such oilseed by-products derived from groundnuts, cottonseed and palm kernel.

Aflatoxin contamination of maize is also an important problem in warm humid regions where *A. flavus* may infect the crop prior to harvest and remain viable during storage.

A. ochraceus produces the ochratoxins. However it shares this property with at least two *Penicillium*species. Ochratoxin A (OTA) and ochratoxin B are two forms that occur naturally as contaminants, with OTA being more ubiquitous, occurring predominantly in cereal grains and in the tissues of animals given contaminated feed. Another mycotoxin, citrinin, often co-occurs with OTA, particularly in wheat.

Fusarium Species

The important *Fusarium* fungi of cereals include: *F. graminearum*, *F. culmorum*, *F. sporotrichioides*, *F. poae* and *F. moniliforme*. These species produce a wide range of mycotoxins. Of particular significance in animal health are the trichothecenes, zearalenone and the fumonisins. The trichothecenes are sub-divided into four basic groups, with types A and B representing the most important members. Type A trichothecenes include T-2 toxin, HT-2 toxin, neosolaniol and diacetoxyscirpenol. Type B trichothecenes include deoxynivalenol (also known as vomitoxin), nivalenol and fusarenon-X. The production of the two types of trichothecenes is characteristic for a particular *Fusarium* species. However, a common feature of the secondary metabolism of these fungi is their ability to synthesise zearalenone which, consequently, occurs as a co-contaminant with certain trichothecenes. The fumonisins are synthesised by a distinct group of *Fusarium* species. Three members of this group (fumonisin B_1, B_2 and B_3) often occur together in maize.

Virtually all the toxigenic species of *Fusarium* described above are also major pathogens of cereal plants, causing diseases such as head blight in wheat and barley and ear rot in maize. Harvested grain from diseased crops is, therefore, likely to be contaminated with the appropriate mycotoxins, and this is supported by ample evidence.

Neotyphodium Coenophialum

The endophytic fungus *Neotyphodium coenophialum* occurs in close association with perennial tall fescue, while another related fungus, *N. lolii*, may be present in perennial ryegrass. Ergopeptine alkaloids, mainly ergovaline, occur in *N. coenophialum*-infected tall fescue, while the indole isoprenoid lolitrem alkaloids, particularly lolitrem B are found in *N. lolii*-infected perennial ryegrass. The ergopeptine alkaloids reduce growth, reproductive performance and milk production in cattle, while the lolitrem compounds induce neurological effects in ruminants.

Phomopsis Leptostromiformis

In Australia, lupin stubble is valued as fodder for sheep, but infection with the fungus *Phomopsis leptostromiformis* is a major limiting factor due to toxicity arising from the production phomopsins by the fungus. Mature or senescing parts of the plant, including stems, pods and seeds are particularly prone to infection. Phomopsin A is considered to be the primary toxin, causing effects such as ill-thrift, liver damage, photosensitization and reduced reproductive performance in sheep.

Pithomyces Chartarum

Pithomyces chartarum is an ubiquitous saprophyte of pastures with the capacity to synthesize sporidesmin A, a compound causing facial eczema and liver damage in sheep.

Use of Organic Acids

Fungal contamination of feeds, particularly high-moisture cereal grains may be prevented by the use of organic acids prior to storage. Propionic acid has emerged as an effective agent for this purpose. However, mixtures of organic acids are now commercially available. The efficacy of BASF mixtures has recently been undertaken in two studies conducted at the Scottish Agricultural College (SAC) and published by Blanchard *et al*. The preservatives tested included Lupro-Mix® NC (containing a mixture of 38 *per cent* propionic acid, 34 *per cent* formic acid, 8 *per cent* ammonia and 20 *per cent* water) and Lupro-Grain® (a mixture of 92 *per cent*, propionic acid, 4 *per cent* ammonia and 4 *per cent* propandiol). The efficacy was compared with propionic acid at similar levels of addition (0.10, 0.25 and 0.50 *per cent*). In the first investigation, the three preservatives were tested against 6 species of storage fungi using a purified medium (potato dextrose agar, PDA) for culture of these moulds. The minimum inhibition concentration (MIC) was then determined for the test products. Propionic acid emerged as the most effective and consistent preservative, with Lupro-Mix® being more effective than Lupro-Grain® in controlling *Penicillium expansum*. In contrast, Lupro-Grain® was more effective than Lupro-Mix® for the suppression of *Aspergillus. flavus* and *A. parasiticus*, while both were equally efficient against *A. ochraceus* and *P. verrucosum* and *P. roquefortii*. *P. verrucosum* was more effectively controlled by the four test compounds than the other storage fungi under the conditions of this investigation.

In the second trial, moist sterilised barley grain was inoculated with A. parasiticus and treated with 0.10, 0.25 and 0.50 *per cent* of the three test products. In contrast to the findings of the first study, Lupro-Grain® emerged as the most effective product with the lowest MIC value. Substrate type influenced the efficacy of test compounds against *A. parasiticus*. Discrepancies may have been caused by differences in water activity (i.e. moisture content) or in buffering capacity of the two substrates. Clearly much more work needs to be done to fully assess the efficacy of these preservatives in other cereals and in compound feeds.

Salmonella Species

Many species of *Salmonella* have been implicated in diseases of farm animals. Of these, *S. typhimurium* is universally distributed while *S. enteriditis* has emerged as a regular pathogen of poultry and contaminant of eggs and chicken meat. Animal feeds are thought to be an important source of these bacteria. Meat and bone meal and fisheal are frequently contaminated with *Salmonella*. Intensive pasture utilisation provides an additional source through contamination of faeces from infected animals. Furthermore, the practice of spreading cattle slurry on to pastures in conventional and organic farms is another potentially significant source of infection.

In many parts of the world poultry manure is used as a feed for ruminants. For example in the USA, two poultry waste products are available for such use: dried poultry waste and dried poultry litter (Jeffrey *et al*., 1998). Dried poultry waste represents undiluted excreta generally derived from caged layer flocks, whereas dried poultry litter is a mixture of excreta and litter. These products are heated to reduce bacterial contamination but are, nevertheless, not sterile. It is reassuring to note that in the trail of Jeffrey *et al*. that all samples of processed poultry litter collected from 13 dairy farms were totally free of *Salmonella* contamination even though virtually all samples contained Enterobacteriacea, non-glucose fermenting Gram-negative and Gram-positive bacteria.

Escherichia Coli

It is widely recognised that cattle feeds contain *E. coli* through contamination with faeces. There is particular concern over the occurrence of *E. coli* O157 since this form has been definitively linked with specific outbreaks of illness in humans.

The application of slurry on to pastures means that there is potential for the transfer of faecal *E. coli* to grazing animals, a practice that has caused some disquiet among those concerned with food safety. Hancock *et al*. (1998) maintain that it is important to distinguish the notion of 'reservoir' from that of an incidental host as regards the source of *E. coli* O157. They were not able to support the thesis that cattle are the reservoir species of this form of the bacterium. The Pennington Group also accepted that impact of spreading slurry on land still requires evaluation but proceeded to recommend the introduction of precautionary measures with this practice. Thus, the contamination of pasture grass with *E. coli* O157 from slurry remains a contentious issue.

Other forms of *E. coli* occur extensively in cattle feeds. Thus, Lynn *et al*. reported that just over 30 *per cent* of cattle feed samples collected from 13 dairies and four feed mills in USA were found to be contaminated with *E. coli*, although *E. coli* O157 was not detected in any of these samples. In five dairies, concentrations of *E. coli* exceeded 1000 colony forming units/g feed. These authors suggested that attention should focus on the replication of *E. coli* in moist feeds and duration of storage in feed bunks. Other studies in the USA also demonstrated the occurrence of *E. coli* (non-O157) in cattle feeds. Jeffrey *et al*. reported that 13 non-O157 E. coli bacteria were isolated from 52 samples of dried poultry litter destined for dairy feeds.

Foot and Mouth Virus

Foot and mouth disease has been a major problem in the UK during 2001/2, with devastating effects on the livestock industry. The disease is spread by inhalation and by ingestion of contaminated materials. The outbreak in the UK has been tentatively attributed to the feeding of catering waste to pigs. While this association may never be confirmed, it is clear that in the intensive rearing of animals due recognition should be given to the need to correctly process animal feeds. However, in the EU, the feeding of catering waste containing meat products is banned under a recent directive. Thus, in theory at least, this source of the foot and mouth virus should be excluded from livestock farms.

Bovine Spongiform Encephalopathy (BSE)

The prion proteins of processed animal proteins have recently emerged as important feed contaminants implicated in the development of BSE in cattle. Prion proteins are normal animal tissue components with the capacity to transform into agents causing fatal neurological syndromes in a wide range of species. The initial onset of BSE was attributed to the feeding of cattle with meat-and-bone meal prepared from carcasses of scrapie-infected sheep. The latter disease is also caused by prion proteins as is the human equivalent, new variant Creutzfeldt-Jakob Disease (vCJD). The incidence of vCJD in humans has been linked with the consumption of BSE-contaminated beef. It is this association that has led to extensive and stringent legislation in the European Union concerning the use of specified animal products in livestock feeding.

Probiotics

A number of microbial products added to the feed may exert beneficial effects on livestock health and production. These products are not attributed with specific nutritional roles and are given the term 'probiotics'. *Lactobacillus acidophilus*, for example, appears to be endowed with the ability to reduce scouring and increase liveweight gain in calves, but the effects are not consistent between trials. Yeast cultures may also be used as probiotics. It should be stressed that the mechanisms of action of probiotics have yet to be elucidated and there is a need for critical cost-benefit analysis of the various products as they become available for commercial use.

Legislation

It might be expected that bacterial contamination of feeds and forages should be comprehensively regulated by statutory measures. However, in the United Kingdom, regulations only exist for *Salmonella* contamination. For example, under UK regulations positive identification of *Salmonella* in feeds must be reported to a 'veterinary officer of the Minister'. Current evidence regarding *E. coli* reservoirs is not sufficiently convincing to warrant prohibition of the spreading of slurry or manure on to pasture. It is difficult to see how faecal sources of this bacterium in ruminant feeds and pastures can be reduced by the introduction of regulations. Any introduction of such legislation will require prior evaluation of the critical control points involved in the initial contamination and subsequent spread of *E. coli* on farms.

The acute toxicity of fungal secondary compounds, and in particular the carcinogenicity of certain mycotoxins has led to the establishment of regulatory and advisory directives for primary feed ingredients and complete rations. Extensive regulations exist for the aflatoxins, but for ochratoxin A and deoxynivalenol only advisory directives have been proposed, unsupported by legislative measures. Of particular concern is the absence of statutory regulations for the carcinogenic fumonisins.

Animal feeds have been subject to additional stringent controls in the aftermath of the BSE crisis in the United Kingdom and elsewhere in Europe. In the UK, it has been illegal to feed ruminants with any form of mammalian protein since November 1994. The feeding of mammalian meat-and-bone meal (mMBM) to any farmed livestock has been prohibited since April 1996. The latter ban was introduced to avoid inadvertent feeding of mMBM to ruminants. Surveillance results of animal feeds indicate widespread compliance with these regulations, with 99.7 *per cent* of feeds found to be negative for prohibited proteins. EU-wide feed directives are also in place regarding BSE control of feeds. Thus, the Council Decision 2000/766 requires Member States to prohibit feeding of processed animal protein to livestock that are bred or destined for food production. The products that are banned under this Decision include mMBM, meat meal, blood meal, dried plasma and other blood products, hydrolysed protein, hoof meal, horn meal, poultry offal meal, feather meal, fish meal, dicalcium phosphate and gelatin. The exceptions to these regulations include the use of milk and milk products and catering waste. However, the use of catering waste that contains or has been in contact with meat or meat products or that originates from premises where the aforementioned foods are handled is prohibited under the Animal By-products Order. Enforcement of the regulations concerning catering waste has come under close scrutiny during the recent foot and mouth epidemic in the UK.

References

Dhand, N.K., Joshi, D.V. & Jand, S.K. 1998. Fungal contaminants of dairy feed and their toxigenicity. *Indian Journal of Animal Sciences*, 68: 1095-1096.

D'Mello, J.P.F. and Macdonald, A.M.C. 1998. Fungal toxins as disease elicitors. *In* J. Rose, ed. *Environmental Toxicology: Current Developments*. Gordon and Breach Science Publishers, Amsterdam, pp. 253-289.

McDonald, P., Henderson, A.R. and Heron, S.J.E. 1991. *The Biochemistry of Silage*. Chalcombe Publications, Marlow, 340pp.

Wallace, R.J and Newbold, C.J. 1992. Probiotics for ruminants. *In* R. Fuller, ed. *Probiotics. The Scientific Basis*. Chapman & Hall, London, pp.317-353.

8

Exposure Assessment of Microbiological Food Hazards

Exposure assessment may be undertaken as part of a risk assessment, or it can be a stand-alone process, such as when there is no information available to undertake a dose-response assessment or when the risk management question only involves quantifying or seeking ways to minimize exposure. Discussions between risk managers and risk assessors may lead to a refinement of the initial question or problem statement to be addressed in the risk assessment, or consultation with other parties may result in the availability of new information, that can in turn lead to revision of assumptions or to further analysis. Exposure assessments are often highly specific to the production, processing and consumption patterns within a country or region.

Principles

The scope of the exposure assessment in terms of content and timeframe should be appropriate to meet its objectives and fulfil the needs of the risk managers. Before embarking on an assessment, the purpose and scope should be clearly identified by those who commission it.

Exposure assessment for microbiological hazards should provide risk managers with a 'best estimate' of exposure that is as free of bias as is possible, along with discussion or analysis of the uncertainties and variability in the estimate. Bias describes forms of error that lead to consistent over-

or underestimation of the true value or average value. The basis of the 'best estimate', whether the average exposure (mean), or the most likely exposure (mode), or level of exposure that is experienced by 95% of consumers, or some other metric, should be clearly communicated, including a description of why that metric is the best measure of exposure. If bias (e.g. the decision to use a worst-case estimate) cannot be eliminated, that bias and the reasons for it should be clearly stated.

Exposure assessment should represent the 'real world' situation as closely as possible and reflect the full range of possible outcomes (i.e. probabilities and levels of exposure), unless risk managers express the need for information on a particular subset of outcomes, such as 'most likely' or 'worst-case' scenarios. It should be noted, however, that deliberately conservative estimates can reduce the usefulness of the estimate for cost-benefit and cost-effectiveness studies, and decrease our ability to describe the uncertainty of the risk estimates. They may be useful in certain situations, however, e.g. to better understand the impact of mitigations.

Specification of uncertainty and variability are critical in terms of correctly understanding and appropriately using the estimate of exposure. It is therefore important to identify these to the greatest extent possible in the exposure assessment, and to discuss their implications on the exposure assessment, and to provide a description of uncertainty and variability as part of the final exposure estimate.

For reasons noted earlier independence and separation of the exposure assessment from the risk management process are highly desirable. Nevertheless, interaction between managers and assessors is also essential to ensure that the exposure assessment provides the best possible support for the decision that the risk-manager has to make, and to ensure that risk managers understand the principles and assumptions underlying the specific exposure assessment.

The need for transparency of the exposure assessment requires full documentation of the process, including sources of data, their evaluation, the models used to assess exposure and any assumptions made, including the effect of those assumptions on the outcome of the exposure assessment.

Purpose and Types of Exposure Assessments

Exposure assessment may be undertaken for different purposes and in different contexts, such as:

- To be combined with a hazard characterization as part of a risk assessment to estimate the risk associated with a pathogen+commodity combination.
- To relate the level of a microbiological hazard in a product to the subsequent potential exposure of consumers. Exposure assessment methods can be applied to foods moving in international trade to assess the equivalence of sanitary measures and demonstrate whether the level of exposure associated with the exported product meets the level of protection required by the importing country.
- To identify where interventions or control options are likely to be most effective in reducing the level of exposure to a microbiological hazard in a given product.
- To compare the efficiency of mitigation measures in reducing the exposure to a given microbiological hazard or to compare the levels of exposure resulting from different processes and food products.
- To compare the exposure resulting from different pathways (cross-contamination vs primary contamination; different contamination sources; different products; etc.).
- To identify information needs and define research activities that could improve the estimation of exposure or control, or both, of the hazard.
- To identify foods in the diet likely to make a major contribution to human exposure to microbiological hazards.
- To evaluate the effectiveness of current protective measures.
- To identify and validate potential Critical Control Points (CCPs) in a process controlled by a Hazard Analysis and Critical Control Point (HACCP) system.

Assessments should be initiated in response to a well-defined risk management question, or questions. If such a question has not been clearly articulated, further discussions with risk managers are needed to define what information is required to support the decisions they have to make and the type of work that needs to be undertaken to provide it. Depending on the risk question, this may include provision of surveillance data, or epidemiological data, through to a qualitative risk assessment or a quantitative production-to-consumption model. Even if a fully quantitative risk assessment is thought to be necessary, it may be useful to commence with a qualitative approach to better define the nature of the work, the feasibility and the time needed to meet the risk manager's requirements.

The questions asked of the exposure assessors, as well as the time, data, information and human resources available, determines the approach (qualitative or quantitative, deterministic or stochastic, etc.) and level of detail necessary, or achievable. The goal of an exposure assessment may be to provide an estimate of the level of exposure to a pathogen in a given population, but may also be limited to evaluation of one or a few processing steps. This again reinforces the need for the risk manager to clearly articulate their needs, the level of detail required in the exposure assessment, and any constraints that would limit the range of management options, to the assessors. For example, when a comparison of potential mitigations is requested, the managers should provide an indication of the measures they would consider or have available for the reduction of exposure from this source as well as any that would not be acceptable under any circumstances.

Purpose and Scope of a Specific Exposure Assessment

Systematic planning is necessary to identify the purpose and scope of the study to be carried out before initiation of the exposure assessment. Time and resources available are also important considerations in the planning stage, and can affect the scope of the work. This planning step is necessary whether the exposure assessment is stand-alone or part of a risk assessment.

The initiation of exposure assessment may be a key point at which the risk profile is revisited. This often requires interaction with the risk managers to ensure that changes in the scope do not reduce the usefulness of the exposure assessment.

The risk profile can include information that assists the exposure assessment. It may include, for example, information on the microbiological hazards of concern, their source or pathway of entry to the food chain, and the difficulties faced in controlling them; an indication of the available data on prevalence and numbers of the microbial hazard of concern in the food chain; disease incidence data and the type and severity of adverse effects; at-risk populations; the foods likely to contribute to consumer perception of the problem; what is expected to be at risk (e.g. human health, economy); and possible control options available. This information assists in defining the risk management question and can also help the assessors determine whether the available data is sufficient to answer the questions posed by the risk managers.

Exposure assessment is normally part of a larger process that includes aspects of setting objectives, communicating findings and managing food safety risks. As noted, before embarking on an assessment, the purpose and scope must be clearly identified by those who commission it. The first step will include a clear description of the food chain, or sections of it, relevant to the particular exposure assessment. The risk manager may also wish to limit the scope to specific regions, or populations, or periods of time. This facilitates the identification of the steps to be modelled within the exposure assessment, which in turn leads to the identification of the data required. In some cases, the necessary data may not be available and it may not be possible to complete the exposure assessment. If this is the case, then communication with the risk manager is required to determine the next step. This could include a revision of the purpose or scope of the assessment or a decision to discontinue the assessment. If sufficient data are available to complete the exposure assessment the next step will be to communicate that conclusion to the risk manager. Exposure assessment is often iterative in nature and involves several stages, including issues and tasks such as identifying data sources and data gaps, deciding on modelling approaches, identifying and assessing the influence of variability, uncertainty and determining sensitivity, as well as assessing the quality of the assessment. All of these will all have to be addressed as part of the exposure assessment process.

Theory into Practice

Once there is a clear understanding of the requirements of the exposure assessment in relation to risk management, the next step is to consider the factors that have a direct effect on consumer exposure to the hazard, including frequency of consumption of the product or commodity; pathway and frequency of contamination with the hazard; the range of doses; and factors that affect it (potential for microbial growth, inactivation during cooking, meal size, seasonal and regional influences, etc.).

In addition, the exposure assessment should describe the relevant pathways of exposure. Scenarios can be constructed to predict the range of possible exposures. For example, if the purpose of the risk assessment is to identify and compare different mitigation strategies to be used from production to consumption, then the entire production-to-consumption pathway has to be addressed. In other cases, only the pathways from retail to consumers may be relevant, thus if the purpose of the exposure assessment were to reach a decision on the maximum tolerable level of a pathogen in a ready-to-eat product at the point of sale, the exposure assessment would be used to determine the potential for further increase or decrease in exposure due to normal consumer handling, such as time and temperature of storage, effect of cooking or other food preparation steps, potential for cross-contamination in the home, etc.

The level of detail required in the different pathways reflects the question asked and the information needed by the risk managers, and may be modified based on the information available. If it has been shown, for instance, that the prevalence or numbers of a specific pathogen differs within a specific commodity according to the type of abattoir, type of processing, type of storage at retail, etc., such information might influence the level of detail required and the selection of pathways in the exposure assessment. Food pathways can be multiple and complex, for example, 'ready-to-eat' meals can be a synthesis of several pathways (meat, vegetable and dressing).

Risk managers may have specific questions concerning specific processes, such as organic farming, logistic slaughtering (i.e. order in which animals are slaughtered) or imported foods that they want to be addressed. Accordingly, these specific interests would need to be taken into account in selecting the pathways to be considered or modelled in the exposure assessment and the types of data to be included.

Resources Required for Exposure Assessment

Some basic capacities are needed to conduct MRA or stand-alone exposure assessment. Risk assessments conducted at the international level (e.g. JEMRA) can assist countries by providing modules or building blocks that can be adapted or modified to suit other exposure or risk assessments; however, exposure assessment in particular usually requires country- or region-specific data. The basic capacities needed include:

- Access to expertise. While the assessment may be carried out by one individual or a small team, access to a range of other expertise usually is needed. Depending on the task, this is likely to include trained risk assessors, modellers, mathematicians, statisticians, microbiologists, food technologists, animal and plant health specialists, agriculture technologists, human and veterinary epidemiologists, public health specialists, and other specialists as identified for specific projects.

- Informed risk managers and policy-makers who are aware of the need for, use of and limitations of risk assessment, working in the context of an appropriate risk management framework, whether in government or industry. This framework must facilitate data collection, decision-making and implementation.
- Financial and human resources to complete the exposure assessment in a timely manner and to an acceptable level that provides useful support for risk management decisions. For very large exposure assessment projects, a dedicated project manager may be desirable.
- Communication channels. Good communication is needed between technical experts, risk managers and the exposure assessors to facilitate efficient exchange of data and knowledge.
- Information technology. Computing facilities, both hardware and software and access to appropriate information networks are needed, to collect, collate and process data, and to provide outputs in a form suitable for communication of results. This should include access to international networks and databases.
- Where data on microbiological hazards are not available, the capacity to conduct surveillance for microbiological hazards, including access to microbiologists, epidemiologists, trained field staff and competent laboratories, is needed.
- While the above list is an ideal, benefits can be also obtained from conducting more modest exposure assessments, but still according to the principles in these Guidelines, even from teams with limited expertise. To assist groups with fewer resources, communication (e.g. including training, mentoring and technology transfer) with more established groups should be actively encouraged.

Modelling Approaches

The goal of exposure assessment is to deduce from the available information the likely probability and magnitude of exposure to the hazard. Detailed exposure data (characterizing the extent of microbiological hazard present in foods at the time of consumption) are usually not available. Thus, exposure assessment will often rely on a model, encompassing knowledge of the factors and their interactions that affect the number and distribution of the hazard in foods, to estimate exposure at consumption.

A model can be defined as 'the description of a system, theory, or phenomenon that accounts for its known or inferred properties and may be used for further study of its characteristics'. Often the model is a simplified description of some more complex system or phenomenon. Models are also used to communicate an understanding, or hypothesis, concerning some aspect of reality that may or may not be able to be directly observed. Thus, another description is that a model is 'an hypothesis or system of beliefs about how a system works or responds to changes in its inputs'. That hypothesis or description can be expressed in words or 'as a system of postulates, data, and inferences presented as a mathematical description of that entity or state of affairs'. A series of mathematical equations that are used to solve a problem (usually involving repetition of one of more operations) can also be called an algorithm.

Among the benefit of a model is that it can be used to predict the outcome of events that have not occurred, or have not been observed. In the context of exposure assessment the models synthesize knowledge from other observations about the pathways of exposure, the behaviour of microbial hazards in foods, patterns of consumption, and so on, to infer what would, or could, happen in other circumstances of interest. Models can be used to interpolate among discrete values of observed data and, in some circumstances, to extrapolate beyond the range of observations. In either case, the validity of the interpolation or extrapolation depends on validation of the model.

Qualitative and Quantitative Exposure Assessment

There is a spectrum of approaches available for exposure assessment, ranging from qualitative to fully quantitative in nature. Qualitative exposure assessments are descriptive or categorical treatments of information, whereas quantitative assessments are mathematical analyses of numerical data. It should be noted that there is a gradation of model types from qualitative to fully quantitative and while such classifications may be helpful, they are not strict defined categories.

A qualitative assessment may be undertaken as part of a first evaluation of a food safety issue to determine if the risk is significant enough to warrant a more detailed analysis but qualitative exposure assessments may, in some circumstances, provide the decision support needed by the risk manager. If a more detailed analysis is warranted, then a fully quantitative assessment is usually the preferred approach if data, time and resources are available to support it.

A literature review or summary of the issues is a valuable first step in the assessment. That literature review or summary should follow the same systematic approach as a quantitative assessment and identify factors that contribute to exposure and how those factors affect the level of exposure. Adopting the structure of the risk assessment process in the creation of the risk profile will facilitate the use of the collected information in any future risk assessment.

Qualitative Exposure Assessment

If the available data are inadequate to develop a numerical estimate of exposure, a qualitative assessment may be developed by assigning descriptive ratings of probability and severity such as 'negligible', 'low', 'medium' or 'high' to the exposure factors. If such an approach is used, specific definitions of the assigned ranges for each rating must be clearly described and justified because 'qualitative' statements and measurements can be misinterpreted. Every effort should be made to refine the assessment so that it can achieve the previously defined goals of exposure assessment.

Potential difficulties with qualitative assessments

Although the resources required for undertaking a qualitative assessment are often much less than those for a quantitative assessment, it is not necessarily a simple process. It is difficult to assign qualitative statements to the data collected. For example, how does one determine if prevalence is 'very low' or 'low'? One person's interpretation of these descriptions may be very different from another's, although the descriptions of the categories would normally be based on expert knowledge. Often elicitation of the knowledge and opinion of experts will be used in qualitative risk assessment.

A second difficulty that can arise in qualitative risk assessment is the combining of qualitative statements at each stage. Careful thought must be given to the types of variable being combined. For example, are two probabilities to be combined, or a probability and a number? Examples of combining the assessed values at different stages might include:

- If the probability of exposure per serving is 'very low' and the level of consumption per year is 'low' then it might be concluded that the probability of exposure per year might be 'extremely low'.
- If the probability of an individual animal being infected with a pathogen is 'high' and the probability of the animal excreting the pathogen (if infected) is also 'high', then the overall probability of any random animal excreting the pathogen might be considered to be 'high'.

It is more difficult to combine the assessed values when one value is high and the other is low, because it may be difficult to assess whether the contribution of one factor outweighs that of the other factor. Then it may be more relevant to assess exposure by considering a range of individual scenarios. In general, qualitative exposure assessments are more reliable at predicting high or low levels of exposure but are much less reliable at assigning intermediate levels of exposure...

Semi-quantitative Exposure Assessment

Semi-quantitative risk assessment provides an intermediary level between the textual evaluation of qualitative risk assessment and the numerical evaluation of quantitative risk assessment by evaluating risks with a score. It offers a more consistent and rigorous approach to assessing and comparing risks and risk management strategies than qualitative risk assessment, and avoids some of the ambiguities that a qualitative risk assessment may produce. It does not require the same mathematical skills of quantitative risk assessment, nor does it require the same amount of data, which means it can be applied to risks and strategies where precise data are missing. Nonetheless, all forms of risk assessment require the greatest possible collation and evaluation of data available on the risk issue, and food safety risk assessments require in-depth knowledge in a variety of scientific disciplines.

Semi-quantitative risk assessment is a relatively new idea in food safety. CAC and others generally consider just two categories of risk assessment: qualitative and quantitative. Semi-quantitative risk assessment, as described here, has often been grouped together with qualitative risk assessment, but this understates the important differences between them in their structure and their relative levels of objectivity, transparency and repeatability.

Quantitative Exposure Assessment

Quantitative exposure assessments provide numerical estimates of exposure, although most models use combinations of mathematics and logic statements. Quantitative exposure assessments require the development of mathematical models in which all relationships between factors affecting exposure can be described mathematically and using logical tests and conditional statements (e.g. 'if' some condition applies 'then' the result is...) in the model.

In a mathematical model, 'input' variables are those that determine the type and magnitude of the response or 'output' variable. The output variable in exposure assessment is the frequency and magnitude of exposure of consumers to the microbiological hazard in the food of interest. In an exposure assessment, input variables would include factors such as time, temperature, production volume and dilution during processing. 'Parameters' quantify the relationship between the input variables and the output(s), and can be fixed values or a distribution. For example, while bacterial growth is proportional to temperature, a scaling parameter is needed to be able to estimate growth rate from the temperature. That value could be fixed for a specific strain of a species, but would vary for different strain of the same species, and in that situation could be described as a distribution.

Quantitative assessments can be divided into two categories: deterministic and stochastic, sometimes also referred to as 'point-estimate' and 'probabilistic' exposure assessments,respectively.

Deterministic

The deterministic or point-estimate approach uses single values such as the average or 'worst-case' for each input in the exposure assessment. The deterministic approach generally requires more data than that needed for qualitative assessment. A single value (average, highest level, most often observed value, 95th percentile, etc.) is chosen to characterize each variable in the model (concentration in the food; the log reduction from cooking; amount of food consumed; etc.). The individual point estimates are combined using mathematical models to generate a point estimate of exposure (worst case, best case, average, etc.). The effects of changes to model variables can then be investigated by 'what-if' testing, and different combinations of variables are used to generate outputs.

When conducting deterministic exposure assessments, selecting a conservative value for each variable has often been used to develop deliberately conservative or 'safe' estimates. Propagating such conservatism through the model can result in an unrealistic over-estimate of exposure because a highly improbable scenario (i.e. the worst possible combination of events) is used to characterise exposure. Thus, a drawback of the deterministic approach is that the likelihood or probability of the estimated exposure actually occurring is unknown. Some values are more likely to occur than others; without knowledge of the likelihood of each outcome, the risk manager may inappropriately allocate valuable resources to reduce an event that rarely occurs. 'Probabilistic' or 'stochastic' models can overcome this problem.

Stochastic

The stochastic or probabilistic assessment represents all the information available for each input variable, described as a probability distribution of possible values. Consequently, the outcome of a stochastic exposure assessment is a statistical distribution that describes both the range of doses of the hazard that might be experienced by an individual or population, and the likelihood of each level of exposure.

The transition from qualitative assessment to deterministic assessment to stochastic assessment usually represents an increase in both information and time required. However, due to the availability of simulation modelling software, the time involved may not be much greater than for an analysis where variable values are summarized as means or specific percentile values. Despite its increased computational complexity over the deterministic approach, much of that complexity is dealt with by the software and the stochastic method is favoured among most risk assessors because it generates more information to support decisions, e.g. by identifying the range of possible exposure levels from all possible exposure routes from which the most likely level of exposure, or any specified percentile value, can be determined. This output provides much greater information than a single point estimate.

Monte Carlo simulation

Stochastic models are generally complex in nature, and as a result are usually difficult or impossible to solve analytically. In fact, even quite simple stochastic models can be impossible to solve analytically. To overcome this problem, the model can be evaluated on a computer, using Monte Carlo simulation. A variety of specialized computer software packages are available to support this approach, and are discussed in various texts. The most commonly used ones are spreadsheet add-ons, such as @RISK® and Crystal Ball®. Microbial risk assessors have also used a stand-alone package called Analytica®. Other mathematical or statistical packages can be also be used for simulation modelling, and other free add-ons are available. Models can also be constructed using general-purpose programming languages, including FORTRAN, Visual BASIC or C. Commercial software packages may be less 'flexible' to use compared to programs developed in programming languages by the modeller. Clearly the latter requires additional expertise and may be more difficult to distribute to others for use or critique. Conversely, commercial Monte Carlo simulation software may facilitate exchange of models, increasing their ability to be 'audited' by others. Simulation models that can be placed and run on the internet may also be desirable to further facilitate model evaluation.

Monte Carlo simulation is essentially an extension of 'what-if'' testing as described for deterministic modelling above. A mathematical model is constructed to describe the exposure assessment pathway including all variables that influence the exposure. Typically this will have the form of a series of linked mathematical equations. Collectively, the result of the combined equations is an expression of consumer exposure. The model is either written in, or transferred into Monte Carlo simulation software, which enables the assessor to easily specify probability distributions, rather than discrete values, for each variable. The software then evaluates the model by selecting, at random, a value for each variable that is drawn from the distribution specified for that variable. The probability of any value being selected from the range defined is in accordance with the probability distribution used to describe that variable. Using the selected values, the values are combined according to the mathematical equations that comprise the exposure assessment model, and the exposure is calculated. This selection and calculation process is called an iteration of the model and represents the exposure from one possible combination of circumstances. There are many such sets of circumstances, however, some more or less likely than others and leading to greater or lesser exposure. To estimate the full range of possible exposures and the likelihood of each, the simulation software repeats the calculations many times: tens of thousand or even millions of iterations are commonly performed. The result of each iteration is recorded and a histogram of the range of exposures and probability of each is generated and forms the exposure assessment.

Other model classification schemes

In addition to the classification of models used in quantitative exposure assessment as deterministic and stochastic, other non-mutually-exclusive classification 'schemes' might be encountered, i.e. the use of one description does not necessarily preclude an additional description from another classification scheme. Several common schemes are mentioned below.

Models can also be categorized as empirical or mechanistic. Empirical models simply describe data or relationships in a convenient mathematical form. Mechanistic models have theoretical bases and, if they are correctly formulated, allow the interpretation of the response in terms of known phenomena and processes. In practice, exposure-assessment models will probably contain both mechanistic and empirical elements.

Estimates of exposure can also be viewed from a temporal perspective: they can be defined as static or dynamic. Static estimates relate to a particular point in time, e.g. the probability and level of exposure associated with a random serving of the food product, or, similarly, the number of contaminated servings consumed per year. In contrast, a dynamic approach would consider the way in which exposure changes over time, for example, reflecting seasonality of exposure or the increasing contamination of a processing line as time from last clean-up increases.

Tiered approach to quantitative risk assessment

Exposure assessments often involve description of very complex systems, where each variable may not contribute equally to exposure and where not all the desired data may be available. In the context of MRA, Van Gerwen et al. suggested that, under such conditions, it could be beneficial to conduct an exposure assessment in a series of stages – they termed this a 'tiered' approach. Similar approaches have been suggested. A rough estimate is first made of the order of magnitude that individual factors or parameters may contribute to exposure or risk. This could be considered as analogous to preparing a risk profile. For those that contribute most significantly, a more detailed assessment is performed, or more data are gathered and combined in, for instance, a deterministic approach. Where relevant, an even higher level of detail can be achieved using, for instance, a stochastic approach.

Van Gerwen et al. propose that, when using a tiered approach, both efforts and resources are focused where they add most to reducing uncertainty in the exposure estimate.

This latter approach may be particularly useful when there is an urgent need for an estimate of exposure.

Modelling the Production-to-consumption Pathway

The way in which exposure estimates are derived depends on the combination of risk management questions being addressed and the amount of data and other resources available, such as expertise and time. An exposure assessment that considers the events from agricultural production through to consumption will demand the most time and resources. Such an exhaustive approach may be appropriate if:

- the risk management questions require consideration of all stages (e.g. the effect of mitigation at the farm, estimates of exposure in final product as consumed), and
- there are sufficient data, knowledge, time and expertise to enable consideration of each stage.

Model development

'Conceptual model' is a term used to describe our understanding of the routes by which the population of interest is exposed to the hazard of concern, including all the factors and their interactions that affect the level and probability of exposure. The conceptual model may be expressed in text or diagrams, or as a mathematical model. Different approaches can be used to develop the conceptual model. The Event Tree approach describes a scenario from a contamination event to a defined end-point of the assessment, e.g. consumption. This approach serves to describe the high-risk pathways that lead to contamination and subsequent disease, and may identify variables in need of further data or modelling. Conversely, the Fault Tree approach begins with the occurrence of a hazard and from there describes the events that must have occurred for the hazard to be present. This approach can provide a framework to analyse the likelihood of an event by determining the complete set of underlying conditions or events that allow the given event to occur.

Additional approaches to modelling used in assessments of microbial food hazards include the Dynamic Flow Tree model and the Process Risk

Model (PRM). The former emphasizes the dynamic nature of bacterial growth and incorporates predictive microbiology using statistical analysis of data, whereas the latter focuses on the integration of predictive microbiology and scenario analysis to provide an assessment of the hygienic characteristics of a manufacturing process.

A more recent general framework is the Modular Process Risk Model (MPRM), which can be thought of as an extension of the PRM approach of Cassin, Paoli and Lammerding. The fundamental assumption of the MPRM approach is that at each of the steps or key activities in the various intermediary stages from production to consumption, at least one of a number of processes can be assigned. These processes can be divided into microbial and product handling processes. The microbial processes include growth and inactivation, and the food and product handling processes include mixing of units, partitioning of units, removal of parts of units and cross-contamination of organisms among units. The transmission of infection among live animals during primary production could be viewed as an additional biological process, which provides the starting estimates of prevalence in a full production-to-consumption model.

When developing mathematical models, the model structure can facilitate or hinder probabilistic analysis and sensitivity analysis. It is recommended that the models should be formulated such that independent variables affecting exposure are clearly specified and in such a way that paired data for each iteration of the model can be stored for all inputs and outputs for which sensitivity analysis is required. When using commercial Monte Carlo simulation software, storage of paired data is automated and users simply need to nominate which variables to 'track'.

The definition of 'unit' is crucial when modelling these processes from production to consumption. A unit is defined as a physically separated quantity of product in the process, for example an animal, a (part of a) carcass, or a package of ground beef, a milk tank or a bottle of milk. It may be that one unit from primary production is also the consumer package (e.g. an egg or whole chicken), but most examples are more complex (e.g. beef carcass to ground beefburger). In this case, units have to be redefined for each stage and at each mixing or partitioning. Both the number of organisms (N) and the prevalence (P) should be treated as uncertain and variable throughout the model. This makes it possible to assess the uncertainty and variability in the final exposure, and thus the uncertainty in the final risk estimate. Of course, knowing N and unit size (U), one could calculate concentration as N/U, which is the number of microbes per unit quantity (g, litre, cm^2, etc.).

It is difficult to suggest a general model framework for cross-contamination; therefore this process is discussed (where relevant) at the individual stages from production to consumption. Useful discussion of this topic can be found in Schaffner.

Variations exist in modelling approaches (e.g., Event trees, Fault trees, Dynamic Flow Trees, PRM, MPRM, etc.). The approach used therefore depends on the perspective of the developer and on the problem being modelled, as indicated by the risk question.

Approaches to modelling the stages from production to consumption are outlined below.

Primary production (farm)

The main focus of the primary production or farm stage of the exposure assessment is to estimate the prevalence and level of the microbiological hazard in the population of interest. For example, this might be prevalence and contamination levels per live cow, per bird, per homestead, apple or per vat of raw milk. One consideration here is to ensure that within the model, for animal products, infection and colonization are differentiated from contamination. These may of course be dependent on each other, such as where excretion by infected or colonized animals may result in contamination of that animal as well as any other animals in the group. Recognizing and incorporating dependencies is an important aspect of constructing robust and logical models.

The level of detail required in the farm model depends on the risk questions being addressed. This detail will relate to whether or not transmission of infection or contamination is included. The model of Hartnett et al., for example, considers transmission on farm while the models of Cassin et al. and USDA/FSIS do not.

Modelling without consideration of transmission of infection or contamination

If investigation of on-farm control is not relevant, estimates of prevalence and numbers of organisms at the point of slaughter or the start of food processing are sufficient. In the case of animal food products, abattoir studies where a number of animals are sampled at the point of slaughter from numerous herds or flocks and throughout the year can be used. This type of information will enable any regional or seasonal variation to be incorporated. In addition, the effects of management factors could be included. Levels of microbiological hazard may be affected by seasonal differences in the prevalence and rate of excretion of microbial pathogens, and by feeding practices and stress. Consideration should also be given to the sample frame and how the sample relates to the population.

Prevalence (national, regional, seasonal, etc.) will be an uncertain parameter within the model. To describe this uncertainty, a number of statistical techniques are available. The Bayesian approach, for example, considers prior information about prevalence and then updates the estimate based on more recently collected data.

The number of organisms on or in an animal and on a product will vary from animal to animal or product to product. To describe this variation, a number of different probability distributions can be used. The choice of distribution will depend on the biological assumptions relating to the contamination process, as well as on the form of the data. Standard texts, such as Vose or Morgan and Henrion, provide comprehensive reviews of these distributions.

Estimating the level of the hazard in the animal or food prior to processing is more difficult than estimating the prevalence. Generally, quantitative data for microbiological hazard is not available or may be limited. Another approach is to look for indicator organisms. Thus, generic E. coli can be used as a measure of the amount of faecal contamination present in a particular food. The amount of faeces present can then be related to the prevalence and level of a particular hazard if the prevalence and numbers of the organism in faeces are known.

If more detailed data are available, a parametric distribution such as the Normal or Lognormal distribution may be fitted to the data and used to describe the variability in numbers. Improved knowledge of the ecology of pathogens in food animals may be needed to help explain the observed variability in the rate of excretion of microbial pathogens. Such knowledge may influence the selection of the most appropriate distribution. Where possible, the results of modelling of hazards in primary production should be validated by available data, e.g. from monitoring and surveillance.

Modelling transmission of infection or contamination

If investigation of on-farm control is pertinent to the risk question, the sources of infection or contamination should first be considered, and subsequently dissemination of the organisms through the population during various production stages should be described. When livestock are considered, these stages include placing of the herd or flock, grow-out and subsequent removal from the facility, and transportation to slaughter. Depending upon the situation in question, it may also be necessary to consider the breeding stages. When the focus is on produce, the on-farm stages consist of planting, cultivation and harvest.

There are numerous factors that will affect the likelihood of initial exposure, including the method of production, environmental factors, physiological factors and human intervention. Moreover, different techniques – such as intensive, organic or free-range regimes – may have distinct impacts upon exposure probabilities. Geographical location, season and altitude may also affect the probability and level of exposure in any given situation. The impact of human intervention in the exposure process should also be considered. It might be, for example, that the continual presence of farm staff increases exposure, whereas the implementation of and compliance with strict biosecurity measures might reduce the likelihood of exposure. In general, information concerning factors that influence likelihood of exposure, such as those described above are, currently, scant or not available in a form for incorporation into risk models. This requires that judgement is required, which may be subjective. Such decisions and their effect should be documented so that their use and basis can be evaluated by users of the exposure assessment.

Once exposed, the prevalence at the point of slaughter or harvest is then a result of the manner in which the organism disseminates throughout the animal group or the crop. Mathematical modelling of this dissemination is an important element of veterinary and plant epidemiology. More details of the techniques used are available. Specific examples used in risk assessment include Hartnett et al. and Turner et al.

Predicting the numbers of organisms per animal or produce unit at the end of the farm stage is more difficult. This will depend on variables that include mode of transmission of infection; excretion rates; growth and survival of the organism in the environment; and other management factors.

Processing

The stages in processing need to be defined before a model can be constructed to describe the changes in prevalence and numbers of organism. The number of stages in the process can be many: Cassin et al., for example, identified 36 distinct processing operations during the slaughter of beef cattle. It is unlikely that all these stages will be followed by all processors, and an added difficulty is elaborating processing scenarios that are both representative of the majority of processors, yet take into account differences in processors. Flow diagrams developed for HACCP systems can be a comprehensive source of information on process steps.

Modelling of processing involves:

- considering the way in which the unit size changes from stage to stage and how this affects prevalence and numbers of organisms;
- considering changes as a result of cross-contamination, without unit size changing; and
- considering changes due to microbial growth or inactivation.

Much effort is expended during food processing operations to minimize microbial growth and/or to maximize microbial inactivation (e.g. using heat), or removal through cleaning and sanitation. Important factors controlling the extent of growth and inactivation are the duration of conditions and severity of treatment (particularly temperature) prevailing during the process.

Studies of the effect of processing on the levels of microbiological hazards are generally limited. Where data are available, they are usually the result of analysis of 'before and after' samples, such as the numbers of organisms contaminating a broiler carcass before and after a stage such as defeathering. The reduction (or increase) in numbers is sometimes modelled using a 'black box' approach whereby the changes are modelled linearly, without attempting to describe any of the underlying microbial processes. Alternatively, mechanisms of recontamination of products in factory environments are discussed in den Aantrekker et al.. Similarly, where changes are due to growth or inactivation, the effects of process duration and conditions on microbial numbers can be estimated using predictive models.

Normally, results from 'before' and 'after' samples are reported in terms of log populations. Caution must be exercised when modelling cross-contamination using log populations. For example, if we assume that a specific cross-contamination event adds 1000 organisms per unit (i.e. log 3) to a unit containing 100 organisms (i.e. log 2) it would be incorrect to simply add log 2 + log 3 = log 5 (or 100 000 organisms per unit). Since the cross-contamination event is independent of the initial number of organisms in the product, the correct calculation involves converting the log counts to their arithmetic value and then adding (100 + 1000 = 1100 (or log 3.04 organisms per unit)). That is, contamination is an additive process. This is in contrast to microbial growth which is a multiplicative process because growth is exponential, i.e. where the log increase is based on the initial number of organisms in the product, and the log values can be added (log 2 initial plus 3 logs growth = 5 logs at the end of growth).

The variation and uncertainty associated with modelling the change in numbers should also be given careful consideration. When choosing the approach, careful thought should be given to what the data represent (variation, uncertainty or both) and how representative it is. For example, a problem with modelling the results of carcass samples is ensuring that the sample region is representative of the entire carcass. One solution is to estimate the magnitude of the bias in a separate study and include this in the model. A practical corollary of this is that if contamination on the carcases is unevenly distributed, when the carcase is broken down into smaller pieces, not all will carry the same level of contamination. This is a good example of the consequence of partitioning and where contamination on each smaller unit may differ. Consequently, the prevalence and number of contaminated sub-units and level of contamination on each sub-unit will need to be described.

Post-processing

The post-food-processing environment includes storage and distribution, retail display and sale, food service operations and home kitchens.

While each of these environments is different, there are many important similarities and some data collected in one environment may be suitable surrogates for conducting exposure assessments in other environments (e.g. contamination or cutting boards), while other data (e.g. storage temperatures) might not.

Complexity

Post-processing environments can be much more complex than processing environments because of the variety of foods involved (restaurant menus, for example, may have dozens of items, and a cafeteria may have hundreds); the complexity of food preparation operations (highly non-linear when compared with food processing operations); differences between operations in terms of physical layout (one kitchen vs another); level of training (new worker vs seasoned veteran); and hours of operation.

The rate of potential growth on contact surfaces can be used to dictate the minimum time interval between successive cleanings of equipment in contact with raw product. Surfaces that become contaminated with films of nutrient-rich liquids from raw product may contain spoilage organisms or bacterial pathogens which could grow on the film. This surface is then replenished with new material from each subsequent unit and can promote cross-contamination to other units. For example, consider that a workshift may be of 4- to 8-hours duration and that the working environment is maintained at 10–15°C (such temperatures are maintained in some food processing operations because at lower temperatures workers became less dextrous and are more likely to have accidents and injuries).

Consumption

To characterize the risk from exposure to microbiological hazards in food, it is necessary to know the amount of food consumed and how often it is consumed. The specific characterization of food consumption patterns used in the MRA depends upon the question to be answered by the assessment, as well as the food consumption data that are available to the risk assessor.

Modelling

When modelling food consumption, it is important for risk assessors to understand the specifics of how the food consumption data set was collected and analysed, and to clearly describe how these data were used in the model, including any assumptions used in arriving at the estimates.

Amount of food consumed

Two important aspects of calculating the amount of food consumed, particularly when using results from food consumption surveys, include:

- the population divisor (i.e. whether the total consumption amount is divided by the total population (amount per capita) or only those who consumed the food (amount per eater); and
- the frequency of consumption (per year, per day or per eating occasion).

Amount per capita vs per eater

The per capita amount is calculated by dividing the total amount of a food by the total number of people in the population. This represents an amount of food that is potentially available for consumption. Per-eater amount is calculated by dividing the total amount of food only by the number of people who actually consumed the food.

For foods that are consumed regularly by the majority of the population (e.g. bread), the per capita and per-eater amounts will be nearly equal. For foods that are consumed less frequently or by fewer individuals (e.g. raw oysters), the per capita and per-eater amounts will be quite different.

Amount per year, per day or per eating occasion

Consumption may be calculated as the amount per year, per day or per eating occasion. Definition of the consumption period is particularly important in MRAs because acute, rather than chronic, exposure is of concern. However, chronic exposure may be relevant for some microbial toxins that are released into foods before consumption.

National food production statistics generally report an amount of food per year. A daily consumption amount may be estimated by dividing the total annual amount by 365. It is not possible to estimate the consumption amount per eating occasion from these data alone. Food production data may overestimate consumption because a proportion of product is not able to be sold, or is not consumed, due to spoilage or loss due to other reasons. For highly perishable products (meat, fish, fruits, salad vegetables, etc.), this may be as high as 20 to 25% of production.

Food consumption surveys of individuals allow much more flexibility in estimating the consumption amount. Survey results are frequently summarized and reported on the basis of daily consumption. If the raw data from the survey are available, it may also be possible to calculate the amount of food consumed per eating occasion or per meal (depends on coding system and questions in the questionnaire). The basis for consumption is particularly important when considering foods that may be consumed more than once on a single day. For example, if a person drinks a 250-ml glass of milk at each of three meals, the amount per meal would be 250 ml, whereas the amount per day would be 750 ml.

When calculating daily food consumption from food consumption survey data, it is important to note whether the amount reflects all days of the survey or only the days on which a food is consumed. As an example, in one study, two days of dietary records were collected for individuals participating in the survey. From those data, consumption could be calculated as consumption on the days the food was actually consumed or the average over two days for which each person participated in the survey. If individuals who ate fish reported consuming 200 g of fish on only one of two days of the survey, the average daily consumption would be 100 g if calculated as the average of all days, or 200 g if calculated as the average amount on days when fish was consumed.

Importance of characterizing the distribution of contamination

The importance of modelling the distribution of the number of organisms in a food will depend on the dose-response relationship for that organism. If a high level of growth occurs in a single unit of food, only one person is likely to become infected because that single unit of food will be consumed by one person. Assuming that there are more than enough cells of the pathogen present to cause infection in most individuals, if that same dose were spread equally over a hundred servings, then the same dose might be enough to infect many of the 100 consumers (assuming a pathogen with a low 'infectious dose'). Conversely, for a pathogen with a very low probability of infection per cell (i.e. a relatively high 'infectious dose'), the predicted risk from the exposure is largely independent of the distribution of doses among units of food. This is because there is, effectively, a direct proportionality between the dose and probability of infection for all realistic doses and for those realistic doses the probability of infection is much less than one. In this particular situation, there is less need to characterize the distribution of the pathogen among different servings. Nauta provides advice on modelling distribution among individual servings.

Consumption frequency

The frequency of consumption may refer to the proportion of the population that consumes a food or how often an individual consumes a food in a specific period. In MRAs, frequency of consumption has been expressed in a variety of ways:

- Number of days per year on which the food is consumed.
- Number of eating occasions over a year:
  - annual number of meals,

  - number of times the food is consumed per year, or
  - number of 100-g portions consumed in a year.
- Percentage of the population who ate the food in a specific period (e.g. a year).

The number of days of consumption during the consumption survey period can be determined directly from the survey results; from that, an annual number of days of consumption may be extrapolated.

The number of meals, eating occasions or individual food items may be calculated directly from the survey results, if the survey covers more than one day per individual. Alternatively, data from single 24-hour recall surveys can be combined with information from food frequency surveys on the proportion of the population who 'usually' consume a food in a given period to estimate the annual number of consumption days.

It may be possible to refine the estimated frequency of consumption by combining food consumption data with other industry information, such as annual sales volume or market share information. For example, if the food consumption data report the frequency of consumption of a broad category such as cheese, market share data may be used to predict the frequency of consuming a particular type of cheese (e.g. Camembert). Note that it might be reasonable to assume that the amount of cheese consumed is similar across types of cheese although the frequency differs by cheese type. As noted above, consideration should be given to the proportion of production that is never consumed due to spoilage, not sold by specified 'use-by' or 'best-before' date, or due to other forms of 'wastage'.

Considerations and challenges in modelling food consumption

There are a number of aspects of food consumption data that should be considered when developing the food consumption model.

Extrapolating data from results of food consumption surveys

Food consumption surveys generally collect information from a subset of the population. If the sample is representative of the total population and statistical weights developed for the survey are used in the data analyses, survey results may be used to predict food consumption patterns for the population as a whole.

For MRAs, it is important to estimate the consumption by sensitive population groups, such as the elderly or the immunocompromised. In the absence of specific data for these groups, it is usually assumed that their consumption patterns are the same as the normal, healthy population of the same age and gender.

Infrequently consumed foods

Estimates of consumption based on a small number of observations (i.e. small number of food consumption records) may be less statistically reliable than estimates based on larger samples. For this reason, care should be taken when interpreting and extrapolating survey results for infrequently consumed foods, even if the survey statistical weights are used in the data analysis.

If the survey data are used to model consumption for an infrequently consumed food, it is important that the consumption amount be calculated from the day or eating occasion on which the food was consumed, rather than as the average of all survey days.

Food consumed as discrete items vs components of mixed dishes

Some foods may be consumed both as discrete items and as components of combination foods or food mixtures. For example, milk may be consumed as a beverage, but also as an ingredient (often in small amounts) in many food items. The normal usage of those foods can also affect hazard levels, e.g. milk consumed in meals may be heated which could reduce pathogen numbers compared to milk consumed as part of a cold milk drink. When modelling food consumption, it is important to know whether the consumption estimate includes all sources of the food or only the amount of food consumed as a discrete item. If the consumption estimate includes the food from all sources, it may be necessary to create a generic 'recipe' for combination foods in order to account for all sources of the food. Alternatively, it may be necessary to estimate from the total consumption only that proportion consumed in a form in which the hazard could be present, such as unpasteurized juice or milk, or hot dogs eaten without reheating. As another example of the effect of partitioning or mixing, while consumption data for shell eggs may indicate that person eats 60g of shell egg per day, in some situations the serving may have been made from many eggs combined, such as scrambled eggs in an institutional setting. In such a case, many consumers might be exposed to a single contaminated egg compared to another situation where a single consumer eats the entire contaminated egg.

Aggregation or grouping of foods

If the risk assessment is focused on food groups rather than individual foods, consideration should be given to the way in which foods are aggregated for estimating consumption. The average consumption amount for a food category is affected by the number of foods it represents and how similar the foods are in terms of the usual amount and frequency of consumption. If the foods are too dissimilar, the average amount and frequency of consumption may be misrepresented. For example, if fluid milk and cheese are grouped together as 'dairy products', the consumption amounts may be quite different and the average consumption will underestimate consumption of milk and overestimate consumption of cheese. Again, if a food category includes seasonal items as well as foods that are available year-round, the frequency of consumption may be overestimated for the seasonal foods. Some consumption surveys do, however, identify seasonal effects, e.g. by sampling individuals at many times throughout the year.

Modelling Microbial Growth and Inactivation

Microbial ecology of foods

The possible responses of most microorganisms in foods include stasis, growth or death. In general, viruses and protozoa ('parasites') are inert in foods, requiring a living 'host' to be able to reproduce. While they cannot grow, they can be inactivated by various treatments and processing steps. Similarly, prions are not infectious organisms but are proteins. While they cannot grow in foods they may be inactivated by some treatments, although they are very resistant to denaturation.

Populations of pathogens in foods may display stasis, growth or death, depending on the formulation of the food ('intrinsic' factors) and the processing, distribution or storage conditions ('extrinsic' factors). Pathogens may even display different responses at different times in a single unit of food because conditions can change during processing, transport, storage and preparation.

While each organism may have a qualitatively similar response to changes in temperature, pH, preservatives, etc., the magnitude and type of response (e.g. growth, death, stasis) to different levels of these factors is specific to the pathogen in question. To estimate exposure at the time of consumption, it may be necessary to model the cumulative effect of the food's composition and processing or storage conditions on the microbiological hazard present. As noted earlier, in some cases, changes in microbial numbers during processing may occur as a result of cross-contamination, rather than growth or inactivation. Note that the same considerations may apply to numbers of microorganisms in water (whether recreational or for drinking and food preparation).

It is important to understand under what circumstances growth, inactivation or cross-contamination may need to be considered. Table 1 gives indicative values, based on expert opinion, for the effect of temperature on the rates of growth or inactivation of many vegetative bacteria. (Inactivation of endospores requires considerably longer times and/or higher temperatures). Growth rates for fungi will be slower, but inactivation rates are generally in the same range.

Table 1. Indicative response times for bacterial growth as a function of temperature.

Temperature (°C)	*Time for 10-fold increase in numbers*	*Time for 10-fold decrease in numbers (hours)*
-80		years to decades
-20		months
0	15–75	
5	10–30	
10	5–20	
20	3–10	
30	2–3	
35	1–2	
50	growth not possible for most	days to weeks
60		hours
70		seconds to minutes
80		fractions of seconds to seconds

Note that each type of microorganism has a finite range of temperature over which it can grow, some preferring somewhat lower temperatures, others somewhat higher temperatures. Note also that the effect of temperature depends on the temperature range considered. At low temperature, survival is enhanced, while at intermediate temperatures, growth at a rate increasing with increased temperature is usual. At high temperatures, however, death results at a rapidly increasing rate with increasing temperature.

Each organism also has a finite range for growth as a function of pH, water activity, organic acid level, preservatives, etc., so that there are upper and lower limits for each factor, as well as an optimal level at which the growth rate is fastest. In general, the inhibitory effects of suboptimal factors interact both to reduce the range of each factor over which growth is possible when one or more factors are suboptimal, and to reduce the overall growth rate. At conditions beyond those that allow growth, stasis – or more probably death – will result at a rate dependent on the conditions.

The growth of microorganisms in a unit of food follows the pattern of a 'batch' culture, often with a period of adjustment ('lag'), involving no growth, followed by exponential growth until some maximum population density (MPD) is reached and population growth ceases. For many organisms and many foods, the MPD is in the range $10^9 – 10^{10}$ cells per gram, ml or cm^2 of food.

Although the ecology of microbiological hazard in food can be complex, predictive microbiology models can be used to estimate changes in pathogen levels in foods as the product moves through the farm-to-fork chain. Ross provides a detailed discussion of the microbial ecology of foods in the context of exposure assessment as part of risk assessment.

Predictive microbiology

For many bacterial pathogens, responses to environmental conditions have been described and summarized in mathematical models that can be used to predict their behaviour in foods, including growth rate, lag time, death rate, probability of growth occurring, and probability of toxin production within the storage life of the product.

In predictive microbiology, foods are characterized in terms of their properties that most affect microbial growth and survival, such as temperature, pH, organic acid levels, salt levels and preservative levels. Microbial responses to analogous conditions are systematically studied and quantified, usually in a simplified laboratory broth model system under static and axenic conditions. The data are collated and summarized as predictive mathematical models.

Conditions actually experienced by the foods and microbes are not static, and the effects of those conditions on rates of growth or inactivation have to be mathematically integrated over time for each of those distinct processes. Thus, measurements of processing and handling parameters, and the duration that these conditions are experienced, are integrated and used to predict changes in hazard levels (i.e. population size or concentration) in the food or water. Some predictive microbiology models, however, recreate the population growth curve, i.e. the number of cells present, assuming a defined starting level, as a function of incubation time. Outputs from such models would normally have to be converted to rates of growth before their application in exposure models.

A potential weakness of many predictive models is that they are developed in laboratory broth media, in which factors such as interactions with other microbes in the food or effects due to the physical structure of some foods and not observed. In general, these limitations relate to a few specific types of products, e.g. lactic acid bacteria may suppress pathogen growth in vacuum-packed or modified-atmosphere packed foods, matrix effects may be important in water in oil emulsions (e.g. butter). While most models have been developed in laboratory broth, models for some microorganisms have been developed in specific foods of particular concern or interest.

Model Types and Availability

Models are available that describe:

- Rates of growth as a function of multiple environmental factors.
- Rates of inactivation, most as a function of a single lethal factor. One should be aware, however, that microbial inactivation is usually considered a stochastic process, i.e. the probability of survival of cells decreases (more or less) exponentially per unit of time. Thus, although the number of viable cells in an individual unit of food may be predicted to be less than one, one might still find survivors if a larger unit of the product (e.g. the total volume of a batch), or many units of the product, were examined or considered.
- Limits to growth as a function of multiple environmental factors (so-called 'growth/no growth' or 'interface models'), though few have yet been published. Absolute limits to growth of many pathogens due to individual environmental variables have been documented by ICMSF.
- Probability of growth or toxigenesis within a defined period as a function of multiple environmental factors.

In addition to numerous small-scale research projects to model microbial responses in foods, two large-scale predictive microbiology research programmes were undertaken in the early 1990s. They were funded by the governments of the United States of America and of the United Kingdom, and resulted in the development of a suite of models for responses of populations of food-borne microbial pathogens and some spoilage organisms. The outcomes of those programmes, and subsequent developments, are now available without cost on the Internet through the Predictive Microbiology Portal and through ComBase Predictor. These software packages include growth models for many pathogens and some spoilage organisms, and a number of inactivation models for some pathogens. ComBase is a database of observations for many published and unpublished sources on microbial growth and inactivation rates, and at the time of writing contains approximately 40 000 records. The database is derived from the USA and UK government-funded research programmes referred to above, from data abstracted from the published literature and from data donated by research organizations around the world.

The integration of models for microbial growth, growth limits or inactivation into unified models that can predict both increases and decreases in microbial populations over time will also improve the utility of predictive models for exposure assessment. Several unified models have been proposed, but none have been widely used or endorsed.

Many reviews of predictive microbiology, including potential pitfalls, have been published. McMeekin et al. and Ross, Baranyi and McMeekin provide a good introduction to the concept and its practical application, and the text edited by McKellar and Lu provides a more contemporary review of the state of the art. An extensive listing of available predictive microbiology models was presented in Ross and Dalgaard. Whiting and Buchanan and Ross discuss the role of predictive microbiology in exposure assessment.

Application of Predictive Microbiology in Exposure Assessment

Pragmatically, two features of a predictive microbiology model are critical to its utility. One is the ability to predict accurately microbial responses under all conditions to which the model applies. Evaluation of this ability is loosely termed 'model validation'. Predictive microbiology is not yet a mature science and many currently available models are incomplete or unvalidated, or both. Thus, exposure modelling should include consideration of the validity and reliability of predictive microbiology models, if used. The second feature is the range of independent variables and variable combinations to which the model applies.

Range of Model Applicability

No predictive models currently in use are fully mechanistic (i.e. derived entirely from fundamental theoretical bases), therefore microbial growth or death cannot be reliably predicted in a food in which the conditions are beyond the range of any individual factor included in the data used to develop the model (i.e. predictions can be made by interpolation only).

Different models have different interpolation regions depending on the experimental design used to develop the model. The determination of the true interpolation region and the consequences of extrapolation were discussed by Baranyi et al.. Those authors concluded that models using a large number of parameters were more prone to unreliability resulting from inadvertent extrapolation, because the predictions of the model often changed dramatically near the limits of the interpolation region.

Inadvertent extrapolation can also occur when using stochastic modelling techniques to describe effects of fluctuating variables. This problem may occur for any factor, but temperature is the factor most likely to fluctuate in most real-world examples. Consideration should be given to truncating the tails of the temperature (and other) distributions used to predict microbial growth or death, if necessary, to match the interpolation range of the predictive microbiology model used. Doing so, however, may change the mean, variance and other properties of the chosen distribution in ways that are unintended. The growth limits for the pathogen of concern, and potential for inactivation (if conditions are beyond those limits) should be considered and included in exposure modelling. Growth/no growth models may assist in this regard, and have been included in some exposure assessment models.

Spoilage Flora

The effect of spoilage bacteria on the shelf life of the product should also be considered. Conditions that lead to rapid growth of pathogens may also lead to rapid microbial spoilage. Contaminated products that are obviously spoiled are less likely to be consumed, and thus not lead to foodborne disease, despite that fact that they contain a microbiological hazard. A fundamental rule of stochastic modelling is that no scenario should be modelled that could not actually occur. Thus, it may be necessary to consider the effect of storage conditions on the shelf life of the product in case unrealistically long times at high temperatures are simulated. This can be implemented by correlating model variables that affect growth (e.g. storage time and temperature). Stochastic modelling texts offer advice on how such correlations can be included in models.

On a related topic, other microorganisms growing in the food can influence the potential growth of pathogens. The final cell density of a pathogenic bacterium can be suppressed when the total concentration of all bacteria in the food reaches stationary phase, a phenomenon that has been termed the 'Jameson Effect' and reported by many authors. In many foods, this effect will not happen before spoilage occurs, but in vacuum-packed or modified-atmosphere packed foods such as processed meats and lightly preserved fish, lactic acid bacteria can reach stationary phase without causing overt spoilage and limit the growth of pathogens within the acceptable shelf life of the product.

Sources of Variability and Uncertainty

In stochastic modelling, it is also important to characterize the magnitude of the variability and its distribution about the mean.

Distribution of response times

Using the limited amount of replicated published data concerning growth rate estimates under varying environmental conditions, Ratkowsky et al. concluded that growth rates became increasingly variable at slower growth rates. Microbial response times or rates are often not normally distributed. Distributions describing growth rate and/or response time variability have been described by Ratkowsky et al., Alber and Schaffner, Zwietering et al., Dalgaard et al. and Ratkowsky et al.. Ratkowsky presented a general relationship between the variance in growth response times and the mean of those responses for a range of possible distribution types.

Sources and magnitude of errors

Model predictions can never perfectly match observations or represent reality. Each step in the model construction process introduces some error as described below.

- Homogeneity error arises because some foods are clearly not homogeneous. Current predictive models do not account for this non-homogeneity of foods.
- Completeness error in predictive models arises because the model is a simplification, and other food effects and microbial ecology effects (structure, competition, etc.) that are difficult to quantify are not included in currently available models.
- Model function error is similar to completeness error, and arises mainly from the compromise made when using empirical models, namely that the model is only an approximation to reality. The systematic overestimation of bacterial exponential growth rate by modified-Gompertz models is an example.
- Measurement error originates from inaccuracy in the limitations in the measurement methods used to collect raw data that are used to estimate the parameters of a model.
- Numerical procedure error includes all errors arising from procedures used for model fitting and evaluation, some of which are only methods of approximation.

As rule of thumb, when constructing a predictive microbiology model from data, each additional variable increases the model prediction error by ±10–15% error. In other words, confidence in the predicted growth rate declines when more variables that affect the growth rate have to be considered. The significance of this for predicted exposure depends on the amount of growth predicted to occur. For a three-factor/variable model the magnitude of the 'error' in terms of growth rate and log number of cells would be 30 to 45%, irrespective of the amount of growth predicted. However, in many situations, probability of infection (and thus risk) is related to the absolute number of cells ingested, not the logarithm of dose. Thus, if one generation of growth (log0.30) were predicted, the error in the predicted number of cells would be ±(log0.30 × [0.3 to 0.45]), i.e. ±23–137% of the estimate. If 10 generations of growth were predicted, the 'error' would be ±(log3.00 × [0.3 to 0.45]) which, in terms of numbers of cells would be ±800–2260%.

Data Used in Exposure Assessment

Two categories of data are necessary for the development of an exposure assessment: firstly, data that, in text format, describe the biological and physical processes as well as the human factors involved, and, secondly, numerical data that allow quantitative estimates of exposure to be calculated. The extent to which numerical data is required will vary from one exposure assessment to another, depending on the defined purpose, scope, modelling approach and details chosen.

The Food Product

The food product should be clearly described to facilitate usage of the associated data in an exposure assessment. If a food is surveyed for contamination, sufficient information on the food, microbiological methods and experimental design, and numerical analyses is needed. For instance, where the scientific literature may report an analysis of a food product such as soft cheese, mention only of the food category or sub-product identity is not sufficient. Rather, a complete description of the food, including salt levels, pH, packaging and other relevant information should be provided. Information on geographical area, season, and the degree to which the data represent all manufacturers, distributors or retailers may be very relevant as well. The number of samples is very important; frequently it is not clear from published data which values are from independent samples and which are replicate samples.

Food factors of importance will include whether the product is fresh or frozen, whether it is sold cooked or uncooked, whether or not it is further processed and the extent to which ingredients are mixed. These factors should be specified to a certain degree within the statement of purpose; however, more detailed data collection may be necessary for the exposure assessment. Such data may refer to the different types of sub-products in a food category (e.g. for a broad category such as soft cheese or fermented meats). Such data may also refer to other factors that may influence the prevalence and/or concentration of hazard in the food (e.g. the extent to which the product and sub-products are domestically produced or imported; the different ingredients added; or other products typically consumed with the product).

Growth, inactivation and survival of microorganisms on, or in, the food product are key parameters when considering exposure. It is therefore important to describe any properties known to affect pathogen growth or survival. While pH, water activity and temperature are the most frequently cited properties and typically have the greatest impact on microbial behaviour, many foods will have additional properties with important consequences. These include the levels of fat, oxygen, phosphates, certain spices, acid anions (especially acetate and lactate), nitrite, ionic and non-ionic humectants (sugars, salts, etc.), and antimicrobials such as benzoate or sorbates. Food structure has also been show to play an important role in influencing microbial behaviour in some foods.

The Food Chain

The food chain consists of all stages from primary production to the consumption (including home, restaurant, foodservice, and/or institutional locations), and thus data relating to each of these stages is required. Using meat processing and distribution as an example, the various stages will include the farm; transport to and holding at a slaughterhouse or processing plant; processing; packaging; storage; distribution and retail; transport to the home; handling; food preparation; and consumption. Some of these stages and processes may vary between producers, retailers and consumers and thus it is important to obtain information to describe this variation. Certain stages or processes may be regulated, for example, with respect to the use of chemicals or additives; such regulation and information on the extent to which they are followed in practice may give relevant data to be collected.

Each stage of the food chain may or may not have an effect on the microbial status of the product, that is, whether or not it is contaminated and if it is contaminated, the numbers of organisms present. Considering growth and survival, the times, temperatures, and other ecological factors such as pH at stages that could facilitate growth or impair survival are important. Particular examples of requirements include the duration of, and temperature during, periods of storage or transport, freezing temperatures, pasteurization times and temperatures and cooking times and temperatures if the commodity is sold cooked, and the addition of ingredients that may alter pH. Data that enable description of the variation in these parameters, for example from producer to producer or day to day, are also important. Often, individual stages in the food chain are considered to be static for a specified period. However, certain conditions, such as temperature, are more likely to be cyclic and data should reflect that. While data may be readily available on thermal inactivation, data on other types of thermal or non-thermal processing that affect microbial growth and survival not as readily available.

Throughout the food chain, many control options are available to minimize the risk of microbiological contamination of the final food product. These may be incorporated in HACCP plans that are specific for each product and manufacturing site, and thus may vary substantially between manufacturers. Data should be collected that describe both the methods of control and the extent to which these vary. Examples include cleansing and disinfection methods and the frequency with which these are undertaken; inactivation methods and their critical limits; any testing of live animals and intermediate or final products, with estimates for sensitivity and specificity; handling practices; and the extent to which cleansing and disinfection is employed.

The potential for microbial cross-contamination within the food processing environment is well established. Data that will give insight into the extent to which this occurs is therefore required. Important areas will include, for instance, the level of contact between live and slaughtered animals or between raw and processed vegetable material, worker hygiene, operating equipment and plant design, and methods of packaging. As before, data defining the extent to which variation is present is also important.

The production of a food commodity is a complex process that may involve several stages of mixing and partitioning. For example, the meat from an individual beef carcass will be partitioned and then perhaps mixed with meat from other beef carcasses to produce a ground beef burger. Partitioning and mixing will influence the microbial status of the product, in terms of both likelihood of contamination and numbers, and thus data that is descriptive of these processes should be collected. Typical requirements will include the extent to which both events occur, the numbers of carcasses or products contributing to a mixed product, and characteristics of products obtained through partitioning (including distributions in quantity and size).

Storage and preparation practices, both in the home and in the catering environment, can influence the level of exposure. In particular, growth or decline may occur during storage prior to preparation if the temperature favours either of these processes; cross-contamination, and subsequent exposure, may occur if adequate hygiene practices are not followed; and cooking of the product will determine the final concentration. To address these issues, data should be accompanied by descriptions of relevant details such as: times of and temperatures during storage; typical handling practices and the potential cross-contamination events that could occur during preparation; the extent to which these events occur and the likely numbers of organisms transferred to different locations within the kitchen; the extent to which consumers are exposed to the organisms that are transferred; and typical cooking times and temperatures.

Data used in the exposure assessment should be collected to represent as closely as possible the conditions and practices prevailing in the different stages of the food chain studied. This will not be possible in all instances, and alternative data may be needed. It is important to clearly describe the rationale for selecting the alternative (surrogate) data and to what extent they do or do not describe the prevailing conditions.

Food handling practices vary from region to region or even within the same country based on ethnicity or local habits. With respect to food handling practices in the homes, restaurants and food service establishments (including street foods), adequate descriptions of likelihood and level of pathogen contamination (related to storage times and holding temperatures of cooked food for sale to the public) are only available occasionally, representing an important data gap.

Microbiological Hazard

As exposure assessment typically requires an estimate of the likelihood and level of contamination of the product by the microbiological hazard concerned, the data needed will describe these parameters and the factors that affect them. Contamination may be determined as a percentage of contaminated samples (the prevalence) and/or the number of microorganisms, e.g. CFU/gram of food. It is important that the detection level and sample size are known, as well as the sensitivity and selectivity (or specificity) of the detection method(s) utilized. The sensitivity is the capability of a method to detect a certain number. A method of too low a sensitivity will result in the occurrence of false negatives. The selectivity (or specificity) is the capability of a method to discriminate between different microbiological hazards. Too low a selectivity may result in the occurrence of false positives. Changing methodologies, particularly improvements in the selective media, need to be evaluated. To allow optimal evaluating of data on prevalence and frequency of contamination, proper descriptions of the details (i.e. year, season, geographical location, country, etc.) should be provided. It should be noted that most pathogen testing is presence/absence testing and thus non-enumerative. There are some pathogens for which tests do not yet exist, so even prevalence data may not be easily obtained. Finally it should be noted that the efficacy of testing is frequently dependent on sample size.

The exposure assessment should commence with starting estimates of the likelihood and level of contamination at the first relevant stage. Considering animal production, for example, these starting estimates describe the likelihood of surface contamination of animals leaving a farm and the number of contaminating organisms. If the assessment commences at a later stage, the starting unit of interest will be the carcass, sub-product or product. The data collected should, ideally, enable any variation in the likelihood and numbers to be fully characterized. Likelihood, for example, may vary between producers, seasons and regions, whereas level of contamination may vary between animals, producers, seasons and regions.

When foods of animal origin are concerned, the infectious or colonization status of the live animals at the farm level may be relevant and, depending on the production process, at later stages, for example, evisceration. Thus, data on the probability of infection or colonization and the level of excretion, as well as the extent to which these vary, will be required. With vegetables or fruits the use of animal manure may have a significant impact on likelihood and level of contamination of raw materials or finished products by the microbiological hazard, and such practices (though not advised) should be specifically recorded.

To investigate mechanisms of on-farm control of foodborne microbiological hazards, it is necessary to understand the dynamics of infection and contamination during primary production. The starting estimates of likelihood and level of contamination will be dependent on these dynamics, and thus data describing the underlying processes are necessary. Typical requirements will relate to: possible sources of infection; the mechanisms of transmission between commodities or animals; management factors, including the use of antimicrobials and vaccination; stocking densities; seasonality of infection or colonization; and regional differences and climatic effects. Similarly, certain practices at primary production may reduce or introduce contamination, and should be considered.

Having collected data to describe the starting levels of exposure, information related to changes in the likelihood and level of contamination throughout the food chain is then necessary. Depending on the particular stage, there may be sufficient information that describes actual changes. Alternatively, it may be necessary to predict changes in likelihood and levels and thus data to enable such predictions will be required. Actual data may also be used to validate some of the model assumptions.

Prediction of changes in the level of contamination on the product throughout the food chain will be based, primarily, on predictive microbiology, and thus on growth and survival models. Information on growth and survival at different levels of temperature, pH, oxygen and salt content may be relevant, as are published or calculated growth rates and D-values. Data on the presence of the organism in the water used at any stage, on machinery, on packaging and on hands may be important. Information on the response of the organism to decontamination and on any protective effects that the food product offers at different stages may also need to be considered.

In planning and executing an exposure assessment, the variability in infectivity or toxigenesis of pathogens should be considered, so that representative data are collected. Obviously there is variability between genera or species, but even within a species, only certain sub-types may be relevant in terms of pathogenesis.

If national data on foodborne pathogens are not systematically collected in a country or region, it may be possible to resort to the use of data from another country for some parts of the exposure assessment model. In that case, the rationale for the choice of country and information on the possible limitations of the data in representing the current situation in the country in question need to be clearly documented.

The maximum contamination levels in a raw ingredient or food at retail can be a sensitive parameter in the exposure assessment, and difficult to evaluate. However, the maximum concentration recorded in a small number of servings may not be representative of the likelihood and level of contamination when it is extrapolated to represent millions of servings.

The physiological and physical state of the microorganism in the food remains a relatively unexplored area. Stress, injury and recovery also affect the initiation of growth. Spores will have a distribution of germination times. Many studies use stationary phase cells grown in a nutrient-rich broth at favourable temperatures, and the predicted lag phase duration represents those conditions; cells that contaminate a food may be in a different physiological state. The extent to which the organism is clustered or aggregated may influence growth, survival and cross-contamination.

The Consumer

The final phase of the exposure assessment is the consumer. Two types of information are used in characterizing dietary exposure to microbiological hazard: information on consumer behaviour and data on food consumption patterns.

Relatively little information exists on food handling practices in the home that affect the safety of foods. Food handling practices vary by geographical region or even within the same country, based, for example, on ethnicity, gender and education. Consumer storage times, extent of cross-contamination, cooking times and temperatures, hot holding temperatures and times, and other data are not generally available. Likewise, little information is available about food handling practices by restaurant and food service operations, including street food, which accounts for an increasingly greater proportion of meals in many countries.

QMRA requires information on the amount of food consumed and the frequency with which the food is consumed. The specific expression or characterization of food consumption patterns used in the MRA depends upon both the question to be answered by the assessment and the food consumption data that are available to the risk assessor. Food consumption patterns will probably differ based on population demographics (age, gender, ethnicity, health status, socio-economic group) and seasonal and regional (both national and international) differences in food availability. Consideration of food consumption patterns for sensitive subpopulations (e.g. young children, pregnant women, the elderly and the immunocompromised) and high-risk consumer behaviour (e.g. consuming unpasteurized dairy products or undercooked or raw meat products) are particularly important. Information that enables estimation of variability in serving size will also be important.

Two types of food consumption data are frequently used for characterizing food consumption patterns for MRAs: food production statistics and food consumption surveys. Other sources of information such as retail food sales or purchase data may be useful in filling data gaps in either food production or food consumption survey data.

Food production statistics provide an estimate of the amount of food commodities available to the total population. Examples of this type of data include the FAO Food Balance Sheets and other national statistics on per capita food production, disappearance or utilization. Because these data are available for most countries and are compiled and reported fairly consistently across countries, they can be useful in conducting exposure assessments at the international level. It is important to note that food production statistics do not consider food that is produced but is never consumed due to spoilage or other losses. Also, food consumption patterns may vary widely within a country and the consumption estimates derived from food balance sheets will not reflect this variability. In Africa, south of the Sahara, for example, the majority of the population live on the land and eat what they produce. National food consumption surveys would be of great value here but they are conducted in relatively few countries worldwide. Some countries have carried out Food Basket studies to describe the amounts and frequency of foods consumed. In countries where household food surveys have been carried out, useful information for exposure assessments might be available. In addition, the use of 'Participatory Epidemiology' methods could be of value in data collection as well. Participatory epidemiology is an emerging field based on the use of participatory techniques for gathering information based on community observations, and traditional oral history.

Food consumption surveys provide detailed information regarding the types and amounts of foods consumed by individuals or households and sometimes also the frequency with which the foods are consumed. These surveys usually include a representative sample of individuals or households, from which consumption for the total population or specific population subgroups can be extrapolated. Since serving size directly affects the numbers of pathogen consumed, these surveys may provide a method to determine a distribution of amounts consumed. Although the surveys are usually short in duration (one or two days to a week for each survey participant or household), they provide detailed information about the types of food consumed, as well as when and where foods are consumed.

It is possible that food consumption data may be available for the 'at risk' group for a specific area. Not all national survey data sets have raw data by time of day and place of consumption as well as a total amount of each food consumed, and even if they do, it is often difficult to extract this type of information and analyse it (e.g. the time of day needs to be clearly defined at the time of the survey, as well as when data are sub-divided for analysis, etc.). It also requires fairly sophisticated software to be able to analyse individual dietary data at this level of detail, as opposed to deriving mean or median population statistics. This is particularly true if all sources of a food are required to be aggregated at an individual person level (e.g. apples from raw apples, apple juice and apple pies). In terms of microbiological risk assessment, this addition of food consumed from different sources also has additional problems as each food source is likely to have a different level of contamination of the hazard due to different food processing and preparation routes.

Food consumption surveys collect information on consumption of thousands of individual foods by a representative sample of individuals nationwide. However, they do not record descriptive information about the foods that may relate to food safety. For example, they may not report whether milk was raw or pasteurized, whether a soft cheese was made from raw milk, whether cooked shrimp were domestically produced or imported, or whether a food was packaged by the processor or at retail. For this information, food sales data from industry, trade associations, retail stores and other sources can be combined with results of food consumption surveys to estimate the frequency with which very specific food products might be consumed. Whenever possible these data should be compared with information from epidemiological studies (case control, cohort or outbreak investigations) to verify or calibrate that food survey capture the actual risk factors.

Data Collection, Selection and Utilization

The objectives and structure of the risk assessment determine specific data needs in order to calculate the desired parameters for the exposure assessment. Most data were previously collected for purposes other than for use in risk assessments, although in some cases new data may be collected. The sections below will discuss data collection (for those situations where this is possible), as well as selection and utilization of existing data (where new data collection is not possible). The sections below will also address appropriate methods to find, select, convert to a common metric, combine, calculate or generate appropriate data to fulfil all of the data needs for the risk assessment.

Data Collection

Exposure assessments require data on a range of characteristics, including microbial growth and survival and the prevalence and enumeration of pathogenic microorganisms. Knowledge of the accuracy, reliability and 'representativeness' of these data is essential. Many of the issues concerning data characteristics and sampling are common to most components of the food chain. In this section, some of the issues surrounding the estimation of prevalence and number of microbiological hazards will be discussed. A similar process would be conducted when estimating any number of variables that might be incorporated into an exposure assessment.

Because of the scope and complexity of the topic, only a summary is presented here. Those contemplating the collection of new information for use in exposure assessments would benefit from consultation with someone trained and experienced in data collection.

The process of obtaining an estimate of the prevalence or the enumeration of microbiological hazards can be represented as a sequence of events:

1. Define the research needs.
2. Identify the reference population and study population and obtain an appropriate sampling frame.
3. Design a sampling scheme and identify the sample population.
4. Collect, transport and analyse appropriate samples.
5. Conduct statistical analysis of the data.

The effect of each step on the final outcome is discussed below.

Defining the research question

While an apparently obvious requirement, the research needs are not always defined. This should be the first stage in the process, as all subsequent decisions depend on correctly specifying the problem. Ideally, the question will be guided by an understanding of the key processes along the food chain continuum that determine the risk to human health. Different questions require different study designs. In order to estimate the prevalence of a microbiological hazard in an animal population, one would ideally sample a representative number of animals from the population. A study aimed at enumerating pathogenic microorganisms might screen samples, so that only contaminated samples are analysed quantitatively.

A precise definition of the hazard and the scope of the study are also required. For example, we may wish to estimate the prevalence of Shiga-like toxin producing E. coli (STEC) or focus the study on E. coli serotype O157:H7. In some cases, little is known about the pathogenicity of particular strains of bacteria (e.g. Campylobacter spp.) and, although it is unlikely that all Campylobacter jejuni are equally pathogenic to humans, more refined estimates of the prevalence of virulent strains is not possible at this time. These factors need to be considered when interpreting the results of the exposure assessment.

Identifying the reference and study population

The reference population is the population for which we wish to generalize our findings (e.g. regional, national or international cattle populations; children or the entire population). The study population is the population from which we wish to draw our sample and is usually defined in terms of space and time. The sampling frame is a list of primary sampling units, for example producers, from which we wish to sample. In national studies, obtaining a sampling frame that is representative can be difficult: sources for determining the sample frame include national census data and producer lists, both of which can be either unavailable, available with reporting restrictions or biased.

Designing a sampling scheme and identifying the sample population

Specific sampling schemes need to be used to obtain unbiased estimates (e.g. simple random, stratified random, cluster sampling), with variability often represented by confidence intervals. These may be adjusted for the lack of independence of responses (e.g. cluster samples), finite population corrections and the sensitivity and specificity of the diagnostic tests. The precision of the estimate will depend on the sample size and this should be determined before the study is undertaken. Sample size can be calculated using both classical and Bayesian approaches after consideration of the required precision, prior belief about prevalence and prior knowledge of test sensitivity. The reliability of prevalence and other parameter estimates may also be affected by the proportion of survey non-responses. The effect of poor compliance in large-scale random surveys is often not considered when estimating uncertainty around parameter estimates. When studying problems with sensitive policy issues, such as the carriage of foodborne pathogens, non-response can be an important consideration.

Collection, transportation and analysis of samples

The way in which the sample is collected can have a bearing on the calculated prevalence or numbers of microorganisms in a sample, or both. For example, rectal sampling of animals by digital retrieval or swab has a number of advantages: the sample is 'fresh' and identifiable to the individual animal. However, this method of sampling raises some issues, such as heterogeneity (clustering) in the distribution of bacteria within the faeces that may result in an over- or underestimation of prevalence and shedding. An alternative is environmental sampling (e.g. sampling faecal pats). This requires less effort, is likely to improve compliance and has no animal welfare implications. Furthermore, if the faeces are freshly voided and samples are taken from several parts of the faecal pat, they are more likely to represent the 'true' prevalence and shedding rates. However, this approach also has a number of disadvantages; true random sampling is in practice very difficult to achieve, and, unless the animal is observed defecating, the samples are neither freshly voided nor identifiable to an individual animal (i.e. the same animal may be 'sampled' twice). Furthermore, if infection alters the rate of defecation (e.g. causes diarrhoea in the animal host), this will result in an over-estimation of prevalence.

Collection, transport and storage of samples may all affect the final estimates of prevalence or number of microorganisms. For example, Campylobacter is susceptible to freezing and UV light. Survival for most microorganisms will generally decline in the period between collection and analysis, assuming sample temperatures are controlled. There are a number of microbiological techniques for maximizing survival, including the use of appropriate transport media, cryo-protective broths and techniques for the recovery of viable but non-culturable bacteria. Alternatively, bacterial growth may occur prior to analysis, resulting in an increased sensitivity of detection or a biased over-estimate of the concentration.

Microbiological tests can be characterized by their sensitivity and specificity. It should be noted that some confusion may be caused by the different definitions used for sensitivity in the laboratory and epidemiological sphere. Diagnostic sensitivity, as it is used in epidemiology refers to the ability of the method to detect a microbiological hazard when it is present in a sample, while laboratory sensitivity is concerned with detection limits.

Standardized and validated tests (AOAC International, ISO, AFNOR, APHA, etc.) are preferred, where the test characteristics are known. In the absence of validation, the test sensitivity and specificity are often unknown and are likely to vary at a number of levels, including between test laboratories, within laboratories (over time) and even between samples. Ideally, an analysis of the sensitivity and specificity of a particular method should be determined by the investigating laboratory at the time of analysis. Documentation for the methods used in an exposure assessment should be accompanied by quantitative information on the accuracy of the tests used, i.e. plating or MPN, unless the methods are widely known.

Selection of Data

It is frequently stated that 'all data are biased'. Nevertheless, data should be as representative as possible of the food, microbial or process parameters being assessed and the population consuming the food. Preferred data generally comes from peer-reviewed publications, followed in importance by non-reviewed or unpublished data (government documents, theses, proceedings, etc.). Some data are not available in the peer-reviewed literature (e.g. consumption data), and it should be remembered that even peer-reviewed data are, in most instances, not collected for the purpose of being used in exposure assessments, and thus may not comply fully with all data requirements or be fully representative for the case at hand. Any biases or limitations in the degree to which data represents any particular point of view should be identified and documented (e.g. funding source). When no data are found, expert opinion will need to be used. Generally, the data should be as close as possible to, or specific to, the requirements of the exposure assessment. For example, if the exposure assessment were to calculate the exposure in a particular country, the preferred data would come from that country. The next choice would be data in that region or an analogous one; the final choice would be from anywhere in the world (keeping in mind the purpose of the risk assessment). Selection criteria should include consideration of factors such as geography, time, microbial strain, methodology, equipment type and design, and population demographics. Food consumption data should provide sufficient detail to allow estimates of consumption of the food(s) of interest per meal or per day. The data should be representative of the total population, and ideally will provide information about subgroups within the population.

Statistical analysis of the data

Microbiological tests are continually refined with the aim of improving sensitivity and specificity. However, these test characteristics are rarely statistically evaluated at the time a study is being conducted. The minimum detection level of a test can be assessed by replicated spiking studies whereby samples are inoculated with known numbers of microorganisms from serial dilutions. However, these studies are not straightforward. Evaluating the number of microorganisms in the original sample used for spiking (and subsequent serial dilutions) is subject to error. Furthermore, it may be difficult to obtain an initial pathogen-free sample and microorganisms in 'spiked' samples may not behave in a similar way to those in 'naturally' contaminated samples. These issues require further study. Specificity can be observed in studies using samples inoculated with microorganisms that are representative of those 'naturally' present in the sample, both in type and number.

Estimates of prevalence or number and the associated uncertainty should be based on a thorough consideration of the study design and accuracy of diagnostic tests. For example, uncertainty is affected by the study design, including the sample size. If the sample is from the environment (e.g. faecal pat sampling), we need to consider multiple samples from the same source, differential defecation rates and an unknown population size. These and other sources of uncertainty, such as test sensitivity, can be incorporated within a Bayesian framework to estimate a posterior distribution of prevalence. Similar approaches can be used to estimate numbers of microorganisms shed in faeces and in surface contamination, with associated uncertainty distributions. The relationship between prevalence and numbers of microorganisms carried can also be explored using empirical and mechanistic models parameterized by field data.

Combining data

Frequently there is a limited amount of representative data and it is often preferable to use all of it. However, decisions need to be made when different data sets have different degrees of applicability and relevance to the parameter being modelled. Suggestions on when and how to combine data for Exposure Assessment can be found in Cullen and Frey. Techniques such as meta-analysis also can be used for the purpose of combining data sets. Bayesian approaches may be useful when considering existing knowledge in the light of recently collected information. In certain situations, using Bayesian techniques allows a better estimate of the parameter to be obtained than if the recent data were used in isolation. When a data set is biased, the data may be adjusted before being combined with other data or used in the risk assessment. An example would be when recent research or methods development demonstrated that data collected by one method consistently underestimated the true parameter value by a known amount.

Weighting is often employed so that data sets considered more relevant have more influence on the derived parameter value. Weighting by the number of samples is frequently used, so that larger studies have more influence. Weights may also be used to reflect the expert's belief in the quality and appropriateness of the data. Older data or data from another geographical area might be used in estimating the parameter value, but be given less weight. The selection of the numerical weighting factors is highly subjective and should be explained for full transparency. Composite data sets may be obtained by averaging, method of moments, or maximum likelihood estimates. Careful examination of the different data sets may facilitate estimates of variation (e.g. different microbial strains used in different studies) or uncertainty (residual errors in statistical analyses). Meta-analysis and mixed-effect models can also evaluate data variation.

To avoid inserting the risk assessor's biases into the parameter values, data should not generally be ignored or deleted. However, certain data sets may clearly be inconsistent with the greater collection of data and knowledge. Comparing the size of the remaining distribution with the divergence of the particular data set may suggest that a particular data set should be excluded. This should be done with caution, as the outlier may indicate another source of variation that is otherwise being overlooked. Criteria for excluding data may be subjective and, when employed, should be fully documented for transparency.

Uncertainty and Variability in Exposure Assessment

One has to deal with uncertainty and variability when conducting an exposure assessment. Uncertainty is the (quantitative) expression of our lack of knowledge. Variability is the heterogeneity of the subjects modelled and includes both stochastic variability (randomness) and inter-individual variability. Uncertainty can be reduced by additional measurement or information, while variability can not.

One of the reasons why uncertainty and variability are easily confused is that both can be expressed as probability distributions. When a model is formulated in a stochastic manner, it is easy to implement the model as a

Monte Carlo simulation using available software. However, when uncertainty and variability are not distinguished, the modelling result may be invalid. It has been shown that in some situations, failure to separate variability and uncertainty can lead to erroneous results, and hence incorrect conclusions.

Variability

Variability is always present in biological systems. It is important to realize variability occurs at many levels. Thus, there may be variability in genotype, strain type, time, place, experimental conditions, etc. It is crucial to define the denominator of the variability (like variability per year and variability per flock), and variability from different sources should not be mixed without careful consideration.

Uncertainty

There are many types of uncertainty in exposure assessment, including process uncertainty, model uncertainty, parameter uncertainty, statistical uncertainty, and even uncertainty in variability.

- Process uncertainty refers to the uncertainty about the relationship between the food chain as documented in the exposure assessment and the processes that take place in reality. For example, rare, undocumented events in food production or consumer behaviour may have a relevant impact on the exposure without being fully considered in the model.
- Model uncertainty comprises both the correctness of the way the complexity of the food chain is simplified, and the correctness of all the submodels that are used in the exposure assessment. To enable the construction of the food chain model, process simplification may be inevitable, but the level to which this is appropriate is subjective, and should be reviewed by experts. Submodels used to describe processes, such as growth during storage at a particular stage, are the choice of the assessor and may be based on the availability of both data and models. As different models may yield different predictions, there will be uncertainty about the appropriateness of a given model.
- Parameter uncertainty incorporates uncertainties dealing with errors resulting from the methods used for parameter estimation, like measurement errors, sampling errors and systematic errors. As part of this, statistical uncertainty is defined as the uncertainty quantified by

applying statistical techniques such as classical statistics or Bayesian analysis. It reflects the uncertainty attending the data, given the model that is used.

Finally, one should consider the uncertainty associated with the variability distributions that are applied in the exposure assessment. A clear and practical way to describe uncertainty of a variable model parameter is to first assign a distribution that represents variability (e.g. Poisson (ë)). Then, a distribution that expresses the uncertainty about the parameter(s) of this variability distribution (e.g. ë follows a Normal (1,0.1) distribution) is chosen.

Many types of uncertainty, for example model uncertainty, are difficult to quantify. In some cases, those uncertainties that can be reasonably quantified (statistical uncertainty) are incorporated in to the exposure assessment, and it is then incorrectly assumed that uncertainty has been quantified. Often this uncertainty is then assumed to be the only source of uncertainty, while others have simply been neglected. Also, it should be realized that uncertainty is always subjective and cannot be validated. However, once additional information is added, the uncertainty may decrease.

Uncertainty and Variability Together

It may be difficult to decide how uncertainty and variability should be separated. When model parameter data from the scientific literature are expressed as a mean value with an associated standard deviation, it may be unclear whether this standard deviation is an expression of variability or uncertainty, or both (e.g. when a growth rate is estimated from a set of growth experiments, it is not clear whether the standard deviation in the growth rate expresses uncertainty or variability). We do not know if the growth rate is actually fixed, but cannot be determined precisely by growth experiments, or varies between the experiments but can be determined precisely. Presumably, the standard deviation expresses both. In an exposure assessment model, it may be important to know which characteristic is represented, and to what extent.

When it is unclear how uncertainty and variability should be separated, there are several possible ways to proceed:

- One might decide not to separate the two, which is simple, but not recommended. Implicitly not separating uncertainty and variability means that the model either solely represents variability or solely uncertainty. If this is the chosen approach, then the choice made between uncertainty and variability should be clearly stated.

- One could test the impact of separation, assuming different ranges of weights for uncertainty and variability and exploring the effect on the end result in several scenarios. This will show how important it is to separate uncertainty and variability in the given situation.
- Alternatively, one might first assume that uncertainty is absent. An assumption of omniscience (pretending that everything is known) results in the remaining probability distributions necessarily describing variability. Once the variability is identified, uncertainty can then be reintroduced.
- Another way to access the potential impact of uncertainty is to identify the variable components, set their uncertain parameters to their expected value and run the model. Then run the model as a 'mixed' model where the uncertain and variable components are simulated together. We can then compare the results of the two models to assess the potential impact of uncertainty and the need or otherwise to separate the two by developing a second order model.
- Cullen and Frey suggest that the relative importance of variability and uncertainty can be assessed by inspection of a two-dimensional simulation result plotted in the form of a cumulative distribution function (CDF) with confidence intervals. The mean CDF is a best estimate of variability.
- Alternatively, Thompson and Graham provide an overview of when to select various probabilistic analysis methods depending upon the policy analysis objectives.

Sensitivity Analysis and Uncertainty Analysis

Uncertainty analysis and sensitivity analyses (sometimes called importance analysis) are two tools available that can be used to inform the risk manager about the outcome of the exposure assessment. More specifically, such analyses of an assessment can be used to facilitate the selection of mitigation strategies, identify control points and focus future research by identifying the key areas where more data should be collected. The identification of research areas may be an iterative process whereby the uncertainty in the results are too large for the manager to make a decision, and consequently an analysis might be initiated to investigate what might be the key sources of uncertainty.

Sensitivity analysis

Sensitivity analysis determines the degree of influence a given input has upon the value of the output.

Sensitivity analysis is instrumental in the identification of points in the process where additional data collection is most useful, where monitoring of critical points in the process is of greatest value, and where mitigation strategies could be most efficient.

At the most basic level, to perform a sensitivity analysis, the inputs to the model are assigned fixed values. In a sequential manner, each of these input values are varied across a predetermined range, and the resulting effect on the output is measured. The extent of the variation in the parameter for which the sensitivity of the output is measured is an important consideration. Readers wishing to learn more about sensitivity analysis should consult Frey and Patil, who review and evaluate 10 sensitivity analysis methods

The range over which the parameter in question is varied should be taken into consideration. Obviously, testing for sensitivity in a range that could not possibly occur would not be informative. However, it should also be kept in mind that this range could alter with changes in technology, formulation or a potential management intervention.

Sensitivity analysis can be used to evaluate the effect of alternative scenarios on the model output. These scenarios could include risk mitigation strategies or hypothetical changes that might not be controllable directly by the risk manager, such as changes in consumption patterns as a result of immigration.

A further outcome of a sensitivity analysis is to identify areas that require the implementation of surveillance strategies. A point in the process upon which the output is very sensitive, but is not a potential control option in a risk mitigation strategy is also of interest. However, monitoring to ensure that it does not change may be an important risk management option, thus ensuring it does not undergo a shift in value resulting in undesirable changes in the output. As stated earlier, time is an important consideration when performing a sensitivity analysis. When efficiency is crucial, it may be desirable to only investigate the sensitivity of those processes in the model that are conceived as options for risk mitigation. However, such an approach may miss areas where monitoring might be a crucial risk management strategy.

Uncertainty analysis

An analysis designed to determine the contribution of the uncertainty associated with an input parameter to the degree of certainty in the estimate of exposure.

An uncertainty analysis, as the name implies, provides insight into the uncertainty associated with the exposure assessment. The uncertainty analysis is designed to focus research or data collection activities that could lower the model output uncertainty most efficiently.

Uncertainty analysis involves testing the effect that our uncertainty in a parameter has on the output. The degree to which the uncertainty in a model output is affected by the uncertainty in the model input can rank these effects.

The primary consideration in an uncertainty analysis is that the magnitude of the uncertainty is considered. It is possible that points in the process that may be quite certain, but which are nevertheless very important and could be overwhelmed by the magnitude of very uncertain parameters. The sensitivity analysis, if it were conducted, would identify this situation. It is thus highly desirable that both analyses are performed.

Assuring Exposure Assessment Quality

The adequacy of the exposure assessment to address the specific questions posed by the risk manager should be assessed prior to its use to support risk management decisions. No standard set of guidelines exists currently for this task, but it is expected that the process of 'quality assurance' of any exposure assessment will consist of three main activities: verification, validation (of several types) and finally peer and public review.

Verification and Validation

Verification essentially consists of assurance that the exposure assessment was developed according to best practices. More specifically, for simulation models, verification is often defined as: "ensuring that the computer program of the computerized model and its implementation are correct", that is, ensuring that the mathematical expressions, definitions of data inputs, etc., and logic of the model are correct and correctly implemented.

The best practices are essentially an agreed-upon set of standards and guidelines for the process of doing microbial exposure assessment, such as those articulated in these Guidelines. It is important to note that following such 'best practices' does not depend upon resources or skills, but merely upon the ability to follow a sequence of steps or procedures. Following a well-designed set of best practices will assure that a microbial exposure assessment outcome is credible. Verification should affirm that the output of the model satisfies the decision-support needs of the risk manager, i.e. that it has fulfilled the purpose as specified by the risk manager.

The validation process determines whether the output of the exposure assessment is sufficiently accurate in absolute terms, and thus can be trusted. Within this context, accuracy is the absence of systematic and random error—commonly known in metrology as trueness and precision, respectively. Somewhat more formally, validation has been defined as 'substantiation that a [computerized] model within its domain of applicability possesses a satisfactory range of accuracy consistent with the intended application of the model'.

In assessing the quality of the exposure assessment, three components should be considered: the data, the model and the documentation, but none of these in isolation would determine the overall validity of the assessment. For example, an assessment that uses imperfect data, but is intended to identify data gaps, could be considered equally as 'valid' as one that uses high quality data and is intended to generate accurate exposure predictions. In this respect, it should be noted that a model can be partially valid, if the overall model cannot be validated but individual components or modules are validated on an individual basis. Also, a model that is validated for one purpose may not be valid for another because validation depends upon the objective of the exposure assessment.

Whatever the state of the data and model, it is essential that the documentation is clear and transparent, fully referenced, and states all relevant assumptions and consequences. An exposure assessment with poor quality documentation should not be considered valid, whatever the quality of the data, the standard of modelling or the fitness for purpose.

The quality of the modelling and documentation components of the exposure assessment are within the control of the risk assessor, but the quality of the data is not. If the quality of the data is found to be inadequate to meet the intended purpose of the exposure assessment, discussion with the risk manager should occur to communicate this 'data insufficiency' and, as appropriate, to seek advice to either alter the purpose of the assessment, change the types of output from the assessment, accept but explicitly communicate the caveats associated with the output, or discontinue the assessment.

Model utility

It is important that the processes under consideration be described in a manner appropriate to the purpose of the exposure assessment. The processes to be described in microbial exposure assessments are often complex, and simplification of these processes may be needed to develop a suitable mathematical description. Any such simplification should still allow an appropriate description of the biological, chemical or physical processes considered. Any impact this may have upon the ability of the assessment to address the issues raised by the risk managers should be recognized and carefully considered. The actual modelling approach used should be implemented at a level appropriate to address the purpose of the assessment. The approaches that may be adopted within an exposure assessment are varied, and include whether the assessment is to be qualitative or quantitative, and the degree of biological and mathematical complexity to be used.

Model accuracy validation

Model accuracy validation can be defined as demonstrating the accuracy of the model for a specified use.

Four aspects of model validation can be recognized :

- Conceptual validation (Is the model biologically plausible?)
- Validation of algorithms (Is the model appropriate?)
- Validation of software code (Is the model implemented correctly?)
- Functional validation (Does the model output make sense?)

Conceptual validation concerns the plausibility of the model: whether the model accurately represents the system under study. Simplification of the underlying process in model steps should be realistic, and the model assumptions should be credible. Usually, conceptual validation is qualitative and is best tested against the opinions of experts with different scientific backgrounds. Experimental or observational evidence in support of the principles and assumptions should be presented and discussed.

Validation of algorithms concerns the translation of model concepts into mathematical formulae. Are the model equations mathematically correct, do the model equations represent the conceptual model, under which conditions are simplifying assumptions justified, what is the effect of the choice of (numerical) methods for model solving on the results, do results from different methods to solve the model agree? A powerful method to evaluate the effects of numerical procedures is to compare the results of different methods to estimate parameter uncertainty (e.g. overlaying parameter samples obtained by Monte Carlo (bootstrap) procedures with likelihood-based confidence intervals).

Validation of software code concerns the implementation of mathematical formulas. Good programming practice (e.g. modularity and full documentation) is essential. Specific points of attention are the effects on the model output of machine precision and software-specific aspects. Evaluation of intermediate outputs is a useful process to ensure correct implementation of a model. As noted above, in other terminology this process may be considered as model verification.

Functional validation concerns checking of the model outputs against independently obtained observations. Ideally, it is evaluated by obtaining pertinent real world data, and performing a statistical comparison of simulated outcomes and observations. This usually requires more detailed information than is available. It may be possible to compare results from exposure assessment studies with independently obtained epidemiological estimates of exposure. Such data cannot validate the exposure assessment per se but may produce valuable insights. Bias in data sources also need to be considered when undertaking functional validation. For example, epidemiological data for sporadic food-borne illness is considered to greatly underestimate the true prevalence for many food-borne pathogens : comparison of the output of an exposure assessment model with epidemiological data would need to consider this bias.

Credibility of results can also be established by demonstrating that different sources of data are consistent with output values. These might include intermediate outputs. Cassin et al. provide a good example of such comparisons. When making such comparisons, the different nature of vehicle, microbiological hazard and processes must be accounted for. It should be noted that if the model output does not agree with the observations, it might not necessarily be that the model is wrong. It may be that the observation itself was influenced by an unknown factor (e.g. microbiological methodological insensitivity) or the underestimation of food-borne illness associated with current epidemiological data, as noted above. There may also be a variety of different influences acting in concert to cause the differences in the results.

Communication during Exposure Assessment

Exposure assessors need to communicate with risk managers to agree on the purpose and scope of the exposure assessment from the very start of the process. Effective communication among assessors, managers and stakeholders is crucial in order to ensure that questions asked by the risk managers can be answered and to ensure that the exposure assessment provides the information needed by the risk managers. A tool in facilitating communication between assessors and managers at the initiation of the process is discussion of the desired model output.

Knowledge of data availability and the various relevant processes will increase during the exposure assessment development process. The initial questions asked by the risk managers often need to be modified during the early stages of the exposure assessment, as information and data limitations become clear. Decisions on the final scope of the assessment and questions to be addressed usually require an iterative process. It may be that the exposure assessment models are able to produce more information than actually defined in the original purpose and scope of the exposure assessment. In such cases, the new possibilities for answering new questions should be communicated to the managers.

The exposure assessors must identify and communicate with all those who have relevant data. This may include the elicitation of expert opinion. It may also include industrial sources, or other sources where confidentiality is likely to be an issue. Issues of confidentiality may restrict data availability, but the establishment of strong links and trusted relationships with those people and organizations that hold such information may help to overcome these problems and increase the likelihood of access to such data.

Managers should be fully informed of the strengths and limitations of the exposure assessment to ensure its best use. To that end, the following are important:

- Exposure assessment results should be presented in an objective manner.
- The results should be presented in a way that makes it possible for people without mathematical and statistical background to understand the essential characteristics of the model.
- All assumptions should be fully acknowledged and their impact thoroughly considered or recognized.
- The exposure assessment should explicitly address sources of variability and sources of uncertainties separately wherever possible.
- Results may consist of a range of exposure estimates based on different data, assumptions and models, rather than presentation of a single exposure estimate.
- Tables and graphs are often more useful than textual descriptions in presentation of quantitative results.
- Requirement for additional data should be explicitly communicated.

Finally, it should be stressed that an exposure assessment model is a tool that can be used more than once. This means that in addition to communicating the results related to the exact purpose and scope of the exposure assessment, it would be worth communicating how the exposure assessment model can be used for future research, or for commissioning specific surveillance studies.

References

Anderson, R.M. & May, R.M. 1991. *Infectious diseases of humans: dynamics and control.* Oxford Science Publications, Oxford, UK.

Cullen, A.C. & Frey, H.C. 1999. *Probabilistic Techniques in Exposure Assessment: A Handbook for Dealing with Variability and Uncertainty in Models and Inputs.* Plenum Press, New York, USA. 335 p.

FAO/WHO. 2003. Hazard characterization for pathogens in food and water: guidelines. [FAO/WHO] *Microbiological Risk Assessment Series,* No. 3.

Ross, T. 1999. Predictive food microbiology models in the meat industry. *Meat and Livestock Australia,* Sydney, Australia. 196 p.

Vose, D. 2000. *Risk Analysis: A quantitative guide.* 2nd ed. John Wiley & Sons, Chichester, England, UK.

Zwietering, M.H. & Hasting, A.P.M. 1997a. M*odelling the hygienic processing of foods – a global process overview.* IChemE Transactions Part C: Food and Bioproducts Processing, 75(C3): 159–167.

Bibliography

Adams MR, Marteau P. 1995. *On the safety of lactic acid bacteria.* Int J Food Micro, 27: 263-264.

Aidoo, K.E., 1986, *Lesser-known fermented plant foods*, Tropical Science, 1986, 26, 249-258.

Anderson, R.M. & May, R.M. 1991. *Infectious diseases of humans: dynamics and control.* Oxford Science Publications, Oxford, UK.

Berry, L. 1988. The pathology of mycotoxins. J. *Pathology,* 154: 301-311.

Bhat, R.V., Sashidhar, R.B., Ramakrishna.Y. & Munshi, K.L. 1989. Outbreak of trichothecene mycotoxicosis associated with consumption of mould damaged wheat products in Kashmir Valley, India. *Lancet,* I: 35-37.

Conway PL, Gorbach SL, Goldin BR. 1987. *Survival of lactic acid bacteria in the human stomach and adhesion to intestinal cells.* J Dairy Sci, 70: 1-12.

Cullen, A.C. & Frey, H.C. 1999. *Probabilistic Techniques in Exposure Assessment: A Handbook for Dealing with Variability and Uncertainty in Models and Inputs.* Plenum Press, New York, USA. 335 p.

D'Mello, J.P.F. and Macdonald, A.M.C. 1998. Fungal toxins as disease elicitors. *In* J. Rose, ed. *Environmental Toxicology: Current Developments.* Gordon and Breach Science Publishers, Amsterdam, pp. 253-289.

Davidson, Alan.,2006. *The Oxford Companion to Food*, 2nd ed, UK: Oxford University Press,

Dhand, N.K., Joshi, D.V. & Jand, S.K. 1998. Fungal contaminants of dairy feed and their toxigenicity. *Indian Journal of Animal Sciences*, 68: 1095-1096.

Diaz. E., 2008. *Microbial Biodegradation: Genomics and Molecular Biology* (1st ed.). Caister Academic Press.

FAO/WHO. 2003. Hazard characterization for pathogens in food and water: guidelines. [FAO/WHO] *Microbiological Risk Assessment Series,* No. 3.

Fellows, P., 1997, *Traditional Foods*, Intermediate Technology Publications, UK.

Gram L and Dalgaard P. 2002. Fish spoilage bacteria—problems and solutions. *Curr Opin Biotechnol.* 13:262–266.

Hennequin C, Kauffmann-Lacroix C, Jobert A, Viard JP, Ricour C, Jacquemin JL, Berche P. 2000. *Possible role of catheters in Saccharomyces boulardii fungemia.* Eur J Clin Microbiol Infect Dis, 19: 16-20.

Hui, Y.H., 1994. *Foodborne Disease Handbook,* M. Dekker.

Jelinek, C.F., Pohland, A.E. & Wood, G.E. 1989. Worldwide occurrence of mycotoxins in foods and feeds (an update. *J. Assoc. Off. Anal. Chem.,* 72: 223-230.

Keen E.C., 2012. "Felix d'Herelle and Our Microbial Future".*Future Microbiology* 7 (12): 1337–1339.

Kordylas, J.M., (1990), *Processing and Preservation of Tropical and Subtropical Foods,* Macmillan, UK.

Loureiro V and Malfeito-Ferreira M. 2003. Spoilage yeasts in the wine industry. Int *J Food Microbiol* 86:23–50.

Madigan M., Martinko J. 2006. *Brock Biology of Microorganisms* (13th ed.). Pearson Education.

McDonald, P., Henderson, A.R. and Heron, S.J.E. 1991. *The Biochemistry of Silage.* Chalcombe Publications, Marlow, 340pp.

Oggioni MR, Pozzi G, Balensin PE, Galieni P, Bigazzi C . 1998. *Recurrent* septicemia in an immunocompromised patient due to probiotic strains of Bacillus *subtilis.* J Clin Microbiol, 36: 325-326.

Ottogalli, G. and Galli, A., 1997, *Fermented Foods in the Past and in the Future*, Annali di Microbiologia ed Enzimologia, Italy.

Palmgren, M.S. & Hayes, A.W. 1987. Aflatoxins in food. In P. Krogh, ed., *Mycotoxins in food,* p. 65-95, London, Academic Press.

Pedersen, C.S. 1979. *Microbiology of Food Fermentation*, AVI Publishing Co, USA

Pitt JI and Hocking AD. 1997. *Fungi and Food Spoilage.* Blackie Academic and Professional, New York.

Ranken, M.D., Kill, R.C. and Baker, C.G.J., 1997, *Food Industries Manual,* Blackie Academic and Professional, UK.

Rhodes, A. and Fletcher, D., 1966, *Principles of Industrial Microbiology*, Pergamon Press, UK

Rodricks, J.V. & Stoloff, L. 1977. Aflatoxin residues from contaminated feed in edible tissues of food-producing animals. *In* J.V. Rodricks, C.W. Hesseltine & M.A. Mehlman, eds., *Mycotoxins and animal health,* p, 67-79, Park Forest Hills, Illinois, Pathotox Publishers Inc.

Rose, A., 1961, *Industrial Microbiology*, Butterworths, UK.

Ross, T. 1999. Predictive food microbiology models in the meat industry. *Meat and Livestock Australia,* Sydney, Australia. 196 p.

Ryan KJ, Ray C.G., 2004. *Sherris Medical Microbiology* (4th ed.). McGraw Hill.

SCAN. 2000. *Report of the Scientific Committee on Animal Nutrition on the Safety of Use of Bacillus Species in Animal Nutrition.* European Commission

Smits GJ and Brul S. 2005. Stress tolerance in fungi—to kill a spoilage yeast. *Curr Opin Biotechnol* 16:225–230.

Steinkraus K H, 1996. *Handbook of Indigenous Fermented Foods*. Marcel Decker Inc, New York.

Tournas VH. 2005. Spoilage of vegetable crops by bacteria and fungi and related health hazards. *Crit Rev Microbiol* 31:33–44.

Vaughn, R.H. 1985. *The Microbiology of Vegetable Fermentation*, in "Microbiology of Fermented Foods", Wood, B.J.B., Elsevier, UK

Vose, D. 2000. *Risk Analysis: A quantitative guide.* 2nd ed. John Wiley & Sons, Chichester, England, UK.

Wallace, R.J and Newbold, C.J. 1992. Probiotics for ruminants. *In* R. Fuller, ed. *Probiotics. The Scientific Basis*. Chapman & Hall, London, pp.317-353.

Zwietering, M.H. & Hasting, A.P.M. 1997a. M*odelling the hygienic processing of foods – a global process overview.* IChemE Transactions Part C: Food and Bioproducts Processing, 75(C3): 159–167.